International Association of Fire Chiefs

International Society of Fire Service Instructors

Fire Service Instructor
Principles and Practice
SECOND EDITION

Forest F. Reeder
Alan E. Joos

Student Workbook

JONES & BARTLETT LEARNING

World Headquarters
Jones & Bartlett Learning
5 Wall Street
Burlington, MA 01803
978-443-5000
info@jblearning.com
www.jblearning.com

National Fire Protection Association
1 Batterymarch Park
Quincy, MA 02169-7471
www.NFPA.org

International Association of Fire Chiefs
4025 Fair Ridge Drive
Fairfax, VA 22033
www.IAFC.org

International Society of Fire Service Instructors
2425 Highway 49 East
Pleasant View, TN 37146
www.isfsi.org

Jones & Bartlett Learning books and products are available through most bookstores and online booksellers. To contact Jones & Bartlett Learning directly, call 800-832-0034, fax 978-443-8000, or visit our website, www.jblearning.com.

Substantial discounts on bulk quantities of Jones & Bartlett Learning publications are available to corporations, professional associations, and other qualified organizations. For details and specific discount information, contact the special sales department at Jones & Bartlett Learning via the above contact information or send an email to specialsales@jblearning.com.

Editorial Credits
Authors: Forest F. Reeder and Alan E. Joos

Production Credits
Chief Executive Officer: Ty Field
President: James Homer
Chief Product Officer: Eduardo Moura
Executive Publisher: Kimberly Brophy
Vice President of Sales, Public Safety Group: Matthew Maniscalco
Director of Sales, Public Safety Group: Patricia Einstein
Executive Acquisitions Editor: William Larkin
Editorial Assistant: Marisa Hines
Production Editor: Cindie Bryan
Senior Marketing Manager: Brian Rooney
VP, Manufacturing and Inventory Control: Therese Connell
Composition: Cenveo Publisher Services
Cover Design: Scott Moden
Cover Image: Courtesy of Tim Olk
Printing and Binding: Edwards Brothers Malloy
Cover Printing: Edwards Brothers Malloy

ISBN: 978-1-4496-8827-1

Copyright © 2014 by Jones & Bartlett Learning, LLC, an Ascend Learning Company, and the National Fire Protection Association®.

All rights reserved. No part of the material protected by this copyright may be reproduced or utilized in any form, electronic or mechanical, including photocopying, recording, or by any information storage and retrieval system, without written permission from the copyright owner.

The content, statements, views, and opinions herein are the sole expression of the respective authors and not that of Jones & Bartlett Learning, LLC. Reference herein to any specific commercial product, process, or service by trade name, trademark, manufacturer, or otherwise does not constitute or imply its endorsement or recommendation by Jones & Bartlett Learning, LLC and such reference shall not be used for advertising or product endorsement purposes. All trademarks displayed are the trademarks of the parties noted herein. *Fire Service Instructor: Principles and Practice, Second Edition, Student Workbook* is an independent publication and has not been authorized, sponsored, or otherwise approved by the owners of the trademarks or service marks referenced in this product.

There may be images in this book that feature models; these models do not necessarily endorse, represent, or participate in the activities represented in the images. Any screenshots in this product are for educational and instructive purposes only. Any individuals and scenarios featured in the case studies throughout this product may be real or fictitious, but are used for instructional purposes only.

The procedures and protocols in this book are based on the most current recommendations of responsible medical sources. The International Association of Fire Chiefs (IAFC), National Fire Protection Association (NFPA®), International Society of Fire Service Instructors (ISFSI), and the publisher, however, make no guarantee as to, and assume no responsibility for, the correctness, sufficiency, or completeness of such information or recommendations. Other or additional safety measures may be required under particular circumstances.

6048

Printed in the United States of America
17 16 15 14 13   10 9 8 7 6 5 4 3 2 1

Contents

Part I: Introduction
Chapter 1: Today's Emergency Services Instructor   2
Chapter 2: Legal Issues   8

Part II: Instructional Delivery
Chapter 3: Methods of Instruction   14
Chapter 4: The Learning Process   20
Chapter 5: Communication Skills   26
Chapter 6: Lesson Plans   30
Chapter 7: The Learning Environment   38
Chapter 8: Technology in Training   42
Chapter 9: Safety During the Learning Process   46

Part III: Evaluation and Testing
Chapter 10: Evaluating the Learning Process   50
Chapter 11: Evaluating the Fire Service Instructor   56

Part IV: Training Program Management and Curriculum Design
Chapter 12: Scheduling and Resource Management   60
Chapter 13: Instructional Curriculum Development   66
Chapter 14: Managing the Evaluation System   70
Chapter 15: Training Program Management   74
Chapter 16: The Learning Process Never Stops   80

Fire Service Instructor II Student Application Package   85
Fire Service Instructor III Student Application Package   103

Answer Key
Chapter 1: Today's Emergency Services Instructor   117
Chapter 2: Legal Issues   119
Chapter 3: Methods of Instruction   120
Chapter 4: The Learning Process   121
Chapter 5: Communication Skills   123
Chapter 6: Lesson Plans   124
Chapter 7: The Learning Environment   126
Chapter 8: Technology in Training   127
Chapter 9: Safety During the Learning Process   128
Chapter 10: Evaluating the Learning Process   129
Chapter 11: Evaluating the Fire Service Instructor   130
Chapter 12: Scheduling and Resource Management   132
Chapter 13: Instructional Curriculum Development   133
Chapter 14: Managing the Evaluation System   134
Chapter 15: Training Program Management   136
Chapter 16: The Learning Process Never Stops   138

Today's Emergency Services Instructor

Workbook Activities

The following activities have been designed to help you. Your instructor may require you to complete some or all of these activities as a regular part of your instructor training program. You are encouraged to complete any activity that your instructor does not assign as a way to enhance your learning in the classroom.

Fire Service Instructor I, II, III

Chapter Review

The following exercises provide an opportunity to refresh your knowledge of this chapter.

Multiple Choice

Read each item carefully, and then select the best response.

_____ 1. In past years, where did fire fighter training typically take place?
  A. On the job
  B. At the fire academy
  C. In trade schools
  D. At individual's homes

_____ 2. Fire service instructors should have experience in which of the following?
  A. Public speaking
  B. The subject matter that they are teaching
  C. Application of material to job skills
  D. All of the above

_____ 3. Which instructor attribute brings excitement to the training environment?
  A. Apathy
  B. Monotone voice
  C. Motivation
  D. Lack of confidence

_____ 4. Which NFPA standard governs the duties, knowledge, and skills of the fire service instructor?
  A. 1001
  B. 1021
  C. 1041
  D. 1500

_____ 5. Job performance duties of the Fire Service Instructor I include which of the following?
  A. Creating lesson plans
  B. Writing performance objectives
  C. Developing testing evaluation instruments
  D. Using instructional media and materials to present prepared lesson plans

CHAPTER 1

_____ 6. What type of diagram is used by an organization to identify the various responsibilities and duties of an individual within a department or work area?
   A. Organizational chart
   B. Flow chart
   C. Venn diagram
   D. Standard operating procedure

_____ 7. Why are fire service instructors considered to be mentors of the fire service?
   A. Because of their position in the department's organizational chart
   B. Because of their job description duties and responsibilities
   C. Because they can observe the development of job skills from recruit to journeyman
   D. Because of the rank designation of the instructor position

_____ 8. Which attribute identifies a proper learning environment for classroom and drill ground activities?
   A. Safe from potential hazards
   B. Distraction free
   C. Environmentally comfortable
   D. All of the above

_____ 9. What term describes a process that provides continuity of the organization and security for the community over time?
   A. Strategic planning
   B. Succession planning
   C. Candidate testing
   D. Mission statement

_____ 10. What type of documentation is issued after a person has acquired the required knowledge in a particular field?
   A. Degree
   B. Certification
   C. Certificate
   D. Award

_____ 11. What type of documentation is issued to a person who attends a learning event that has no testing or performance requirement?
   A. Degree
   B. Certification
   C. Certificate
   D. Award

_____ 12. What process helps the instructor stay current on trends and techniques needed throughout his or her career?
   A. Continuing education
   B. Recordkeeping
   C. Certification
   D. Accountability

_____ 13. What/who determines the training priorities for a response agency?
   A. The fire chief
   B. State statutes
   C. Certification or recertification requirements
   D. All of the above

_____ 14. Which of the following methods can fire service instructors use to help in avoiding "burnout"?
   A. Delegate duties to other qualified instructors.
   B. Limit what is taught to make it easy on themselves.
   C. Hire part-time instructors to teach areas they don't want to teach.
   D. Coordinate all training to allow more preparation time.

Matching

Match the following roles or responsibilities with the correct level of instructor responsibility.

A. Instructor I
B. Instructor II
C. Instructor III

_____ 1. Manages instructional resources, staff, facilities, and records and reports
_____ 2. Develops student evaluation instruments
_____ 3. Adjusts presentations to students' different learning styles
_____ 4. Administers and grades student evaluation instruments
_____ 5. Develops a course evaluation plan
_____ 6. Schedules instructional sessions
_____ 7. Reviews and adapts prepared instructional materials
_____ 8. Develops instructional material for specific topics
_____ 9. Plans, develops, and implements curricula

Match the following fire service instructor roles with the correct definition.

A. Leader
B. Mentor
C. Coach
D. Evaluator
E. Teacher

_____ 10. Identifies future leaders of the organization by observing raw talent and ongoing growth
_____ 11. Prepares crews and members for operations by building team skills through practice and repetition
_____ 12. Sets the example for all fire fighters to follow in terms of performance excellence
_____ 13. Presents new skills and abilities in a variety of learning environments
_____ 14. Identifies the level of proficiency that an individual possesses after learning has taken place and provides feedback on performance

True/False

If you believe the statement to be more true than false, write the letter T in the space provided. If you believe the statement to be more false than true, write the letter F.

_____ 1. A Fire Service Instructor I prepares lesson plans for use in fire service training.

_____ 2. A Fire Service Instructor II has many diverse roles and responsibilities in the management of instructional resources.

_____ 3. The Fire Service Instructor II must also perform all of the duties of the Instructor I.

_____ 4. The classroom and drill ground should provide a safe, comfortable, and distraction-free environment whenever possible.

_____ 5. There are many physical factors such as seating arrangement that can both positively and negatively impact the success of the learning environment.

_____ 6. Good instructors and students alike continually place themselves in learning situations.

_____ 7. Training programs are immune from national standards.

_____ 8. Ethics are not typically an issue that an instructor must deal with in instruction.

_____ 9. There are many issues of confidentiality that the instructor will have to deal with, including information regarding learning disabilities.

_____ 10. Accountability in the fire service applies only to fire ground activity.

Fill-in

Read each item carefully, and then complete the statement by filling in the missing word(s).

1. The _____ _____ _____ for instructors are documented in NFPA 1041.

2. As a fire service instructor, you must be aware that your _____ or _____ rests with the degree of effort and preparation put forth at the beginning of your career.

3. _____ set the bar for fire service instructors and are intended to maintain a high level of proficiency and knowledge.

4. _____ _____ _____ spell out what is acceptable and what is unacceptable within a fire department.

5. All training sessions must be _____ _____, along with a factual listing of the objectives that were accomplished during the training session.

Short Answer

Complete this section with short written answers using the space provided.

1. Define the roles and responsibilities of the Fire Service Instructor I.

_____
_____
_____
_____

2. Define the roles and responsibilities of the Fire Service Instructor II.

   _____
   _____
   _____
   _____

3. Define the roles and responsibilities of the Fire Service Instructor III.

   _____
   _____
   _____
   _____

4. Describe the physical and emotional elements of a classroom.

   _____
   _____
   _____
   _____

5. Identify and explain the five major roles of the fire service instructor.

   _____
   _____
   _____
   _____

6. Identify and explain four issues of ethics for fire service instructors.

   _____
   _____
   _____
   _____

Instructor Applications

Discuss your response to these instructor applications to assist you in developing knowledge about your responsibilities as a fire service instructor.

1. List the changes in training delivery that you have seen or experienced since you began your career in the fire service.

   _____
   _____
   _____
   _____

2. Select an industry Web site or reference materials, and identify at least three hot topics facing instructors today.

   _____
   _____
   _____
   _____
   _____

3. Identify a new piece of equipment purchased by your department, and describe how initial training was accomplished.

   _____
   _____
   _____
   _____
   _____

4. Ask the newest or youngest member of your organization to describe how he or she was educated at the last school that he or she attended. Identify similarities and differences between how fire service training and traditional education are presented.

   _____
   _____
   _____
   _____
   _____

Legal Issues

Workbook Activities

The following activities have been designed to help you. Your instructor may require you to complete some or all of these activities as a regular part of your instructor training program. You are encouraged to complete any activity that your instructor does not assign as a way to enhance your learning in the classroom.

Fire Service Instructor I, II, III

Chapter Review

The following exercises provide an opportunity to refresh your knowledge of this chapter.

Multiple Choice

Read each item carefully, and then select the best response.

_____ 1. Which of the following sources of law sets expectations for and restrictions on the fire service instructor's conduct?
   A. Individual fire department policy
   B. Federal law
   C. State law
   D. All of the above

_____ 2. According to the Americans with Disabilities Act, which of the following would be defined as an impairment that is covered by this act?
   A. Inability to swim
   B. Inability to don an SCBA in 60 seconds after repeated training
   C. A hearing or speaking disorder
   D. Compulsive gambling

_____ 3. What term describes an employer's efforts to assist an individual with a disability covered by the Americans with Disabilities Act?
   A. Reasonable accommodation
   B. Physical therapy
   C. Undue hardship
   D. Eligibility determination

_____ 4. What does Title VII prohibit in the event that an individual files a charge of discrimination?
   A. Compensation
   B. Retaliation
   C. Discrimination
   D. Aggressive behavior

_____ 5. What is the simplest form of responsibility that also places an assignment of blame on an individual or an organization?
   A. Liability
   B. Malfeasance
   C. Misfeasance
   D. Tort

CHAPTER 2

_____ 6. If an organization passes a rule stating it is an "English-only" employer, it could be found guilty of what type of discrimination?
  A. Civil rights
  B. Age
  C. Religious
  D. National origin

_____ 7. Which type of wrong can be alleged in civil cases when an injury occurs that may be equated to the term accident?
  A. Negligence
  B. Malfeasance
  C. Responsibility
  D. Tort liability

_____ 8. One of the first lines of defense in the event that a lawsuit is filed is:
  A. proper recordkeeping.
  B. purchasing liability insurance.
  C. having a witness for everything you do.
  D. citing case studies when presenting information.

_____ 9. The best practice for an organization is to put what type of policy in place to prevent discrimination?
  A. A sexual harassment policy
  B. An Americans with Disabilities Act compliance policy
  C. An antidiscrimination policy
  D. A policy of not hiring protected class individuals

_____ 10. What type of files and records must be kept secure because they may be protected by state and federal privacy laws?
  A. Personnel files
  B. Hiring files
  C. Disciplinary files
  D. All of the above

Matching

Match the following types of law with the correct definition.

A. Standards
B. Statutes
C. Codes/regulations

_____ 1. Created by legislative action and can be either state or federal

_____ 2. Any rule, guideline, or practice recognized as being established by authority or using common language

_____ 3. Created by an administrative agency with the authority to do so

Match the following type of law with the example that best identifies it.

A. Standard
B. Statute
C. Code/regulation

_____ 4. Building code
_____ 5. NFPA 1403
_____ 6. Americans with Disabilities Act

Match the following federal employment law with the brief summary of its content.

A. Age Discrimination in Employment Act
B. Equal Pay Act
C. Americans with Disabilities Act
D. Section 1983 of the Civil Rights Act of 1984
E. Title VII of the Civil Rights Act

_____ 7. Prohibits discrimination against a qualified person because of a disability
_____ 8. Prohibits discrimination based on race, gender, or national origin
_____ 9. Requires that all persons regardless of gender be paid at the same rate when performing the same job
_____ 10. Prohibits discrimination that creates a hostile work environment
_____ 11. Prohibits discrimination against persons older than the age of 40 years

Match the type of records or reports with the typical contents of those files.

A. Personnel Files
B. Hiring Files
C. Disciplinary Files

_____ 12. Reports of reprimands or punishment
_____ 13. Individual test scores, pre-employment physical exams
_____ 14. Birth certificate, social security number, and dependent information

True/False

If you believe the statement to be more true than false, write the letter T in the space provided. If you believe the statement to be more false than true, write the letter F.

_____ 1. Every fire department must have a policy in place prohibiting sexual harassment in the workplace.
_____ 2. In a sexual harassment situation, parties must be of the opposite gender.
_____ 3. Many materials you may want to use for training purposes are covered by copyright.
_____ 4. Fire service instructors who joke with students regarding religion or nationality are not liable for those comments because they are known to use comedy as part of their presentation style.
_____ 5. Training is part of an organization's risk management program.
_____ 6. Proper ethical conduct should be the goal of all those involved in the fire service.
_____ 7. Gross negligence is defined as an act that if not intentional shows utter indifference or conscious disregard for the safety of others.

_____ 8. Failing to follow a written protocol that creates a substantial risk to another person can be termed as willful and wanton conduct.

_____ 9. It is a best practice to seek written permission to use copies of written materials for classes.

_____ 10. It would be a violation of copyright law to purchase one copy of a textbook and make copies of it for other students in the class.

Fill-in
Read each item carefully, and then complete the statement by filling in the missing word(s).

1. Race and color harassment includes _____ _____, jokes, and derogatory or offensive comments.

2. The Americans with Disabilities Act defines _____ _____ as a modification or adjustment to a job or work environment that will enable a qualified applicant or employee to perform the job.

3. Sexual harassment consists of _____ _____, requests for sexual favors, and other verbal or physical contact of a sexual nature.

4. Simply defined, _____ consists of improper or wrongful performance of a lawful act without intent through mistake or carelessness.

5. Failure to abide by the _____ provisions required by law or your department could lead to a liability or adverse action against you.

6. Examples of _____ _____ include Facebook, Twitter, LinkedIn, MySpace, and a variety of other Web sites.

Short Answer
Complete this section with short written answers using the space provided.

1. Describe how to document injury to a participant or observer during a training session.

   _____
   _____
   _____
   _____

2. Describe the training records and reports required by the fire department.

   _____
   _____
   _____
   _____

3. Define the types of laws that apply to the fire service instructor.

   _____
   _____
   _____
   _____

4. List examples of the types of disabilities that are covered by the Americans with Disabilities Act.

_____
_____
_____
_____

5. List examples of the types of materials that are covered by copyright laws.

_____
_____
_____
_____

Instructor Applications

Discuss your response to these instructor applications to assist you in developing knowledge about your responsibilities as a fire service instructor.

1. Review your training policy for requirements in documentation of training sessions and recordkeeping.

_____
_____
_____
_____
_____
_____
_____

2. List the local ordinances, laws, and standards that apply to the development, delivery, and recordkeeping of training.

_____
_____
_____
_____
_____
_____
_____

3. Identify the written policies and procedures in place in your organization that address the laws and standards that are discussed in this chapter.

_____
_____
_____
_____
_____
_____

Methods of Instruction

Workbook Activities

The following activities have been designed to help you. Your instructor may require you to complete some or all of these activities as a regular part of your instructor training program. You are encouraged to complete any activity that your instructor does not assign as a way to enhance your learning in the classroom.

Fire Service Instructor I, II

Chapter Review

The following exercises provide an opportunity for you to refresh your knowledge of this chapter.

Multiple Choice

Read each item carefully, and then select the best response.

_____ 1. What factors combine to determine performance in training?
   A. Learning and motivation
   B. Cognitive and psychomotor factors
   C. Time and distance
   D. Interaction and participation

_____ 2. How many types of adult learners are identified in this chapter?
   A. 1
   B. 2
   C. 3
   D. 4

_____ 3. Which of the following is NOT a type of adult learner?
   A. Self-centered
   B. Goal-oriented
   C. Learner-oriented
   D. Activity-oriented

_____ 4. One of the primary questions that affects how you design your class is answered by asking which question?
   A. When will the class end?
   B. Why is the adult learner in the class?
   C. What will the student learn by attending?
   D. How does the classroom have to be arranged?

_____ 5. One attribute of an effective instructor that may be contagious in the classroom that ultimately improves the learning environment is called:
   A. learning style.
   B. empathy.
   C. voice and eye contact.
   D. motivation.

CHAPTER 3

_____ 6. Each adult learner has a distinct and preferred way of perceiving, organizing, and retaining experiences. This is called
   A. motivation.
   B. learning style.
   C. thought process.
   D. perception.

_____ 7. Which generation of adult learners is characterized as the senior membership of most departments and usually offers many experiences and much knowledge?
   A. Generation X
   B. Generation Y
   C. Baby boomers
   D. Generation Z

_____ 8. Which generation of adult learners is characterized as one that is conditioned to expect immediate gratification, that is ambitious, and who tends to include independent problem solvers?
   A. Generation X
   B. Generation Y
   C. Baby boomers
   D. Generation Z

_____ 9. Which generation of adult learners is known for holding a wide range of opinions on a wide range of issues and may be considered pampered and self-centered?
   A. Generation X
   B. Generation Y
   C. Baby boomers
   D. Generation Z

_____ 10. What type of disruptive student is preoccupied with anything other than the lesson being presented?
   A. Monopolizer
   B. Expert
   C. Day dreamer
   D. Historian

_____ 11. What type of disruptive student may become confrontational in order to show off their brain power or perceived knowledge of a particular subject?
   A. Monopolizer
   B. Expert
   C. Day dreamer
   D. Historian

_____ 12. When dealing with disruptive behavior during instruction, one of the most important rules to remember is that:
   A. the student's needs come first.
   B. you are in charge of the classroom.
   C. unproductive behavior must be addressed after the class is over.
   D. everyone has a different learning style, and some people will simply act out.

_____ 13. Which learning process is teacher-directed instruction?
   A. Andragogy
   B. Pedagogy
   C. Formal
   D. Informal

_____ 14. Using Laskowski's acronym AUDIENCE, what does the D identify?
   A. Density
   B. Domain
   C. Demographics
   D. Department

_____ 15. A case study or role-play is an example of what type of leaning?
   A. Structured
   B. Blended
   C. Asynchronous
   D. Enhanced instructional model

Matching

Effective adult learning takes place in a particular sequence. Match the terms with the correct steps of the sequence.

A. First
B. Second
C. Third
D. Fourth

_____ 1. Adult learners use the new content, practicing how it can be applied to real life.

_____ 2. Seeing that "newly" learned information has real-life connections prepares the adult learner for the next step—learning something new.

_____ 3. Adult learners take the material learned in the classroom and apply it to the real world, including situations that were not always covered in the original lesson.

_____ 4. Adult learners begin with what they already know, feel, or need. In other words, learning does not take place in a vacuum.

Match the type of adult learner with the way that he or she learns new information.

A. Learner-oriented
B. Goal-oriented
C. Activity-oriented

_____ 5. Acquires knowledge with the goal of improving job skills or learning a new skill

_____ 6. Requires personal protective time with control over content and learning style

_____ 7. Likes to be involved in the learning

Match the percentage of what we remember to how we receive information.

A. 10 percent   _____ 8. What we see and hear
B. 20 percent   _____ 9. What we read
C. 30 percent   _____ 10. What we say while doing
D. 50 percent   _____ 11. What we hear
E. 70 percent   _____ 12. What we say
F. 90 percent   _____ 13. What we see

True/False

If you believe the statement to be more true than false, write the letter T in the space provided. If you believe the statement to be more false than true, write the letter F.

_____ 1. The type of learner you are affects the way learning takes place.
_____ 2. The monopolizer is a disruptive student who has some experience and wants to make sure everyone knows it.
_____ 3. Behavioral problems displayed by disruptive students may be rooted in other problems.
_____ 4. Calling students by their name during instruction makes them feel important and may get them more involved in learning.
_____ 5. A generation X learner represents a change in learning methods that may involve the use of technology more than others.
_____ 6. Adult learners prefer to have goals set for them in their learning environment.
_____ 7. In the adult learning environment, the adult likes to begin with what they already know or to start from their comfort zone.
_____ 8. As a fire service instructor, you have the responsibility to ensure that the learning environment is appropriate for all of your students, even those not in the mainstream.
_____ 9. Adult learning does not need to be problem or experience centered to be effective.
_____ 10. We remember 90 percent of what we say while doing.
_____ 11. Andragogy is an attempt to identify the way in which adults learn.
_____ 12. All adult learners learn alike.

Fill-in

Read each item carefully, and then complete the statement by filling in the missing word(s).

1. Learning is the potential behavior, whereas _____ is the behavior activator.

2. The term _____ _____ has been used to describe people whose teenaged years were touched by the 1980s and who were the first generation in which both parents typically worked.

3. A(n) _____ _____ can be used to signal an important point in your lecture.

4. We remember only _____ _____ of what we read, but remember _____ _____ of what we say while doing.

5. Factors such as participation in _____ _____ since high school will affect the adult learner's ability to learn.

Short Answer

Complete this section with short written answers using the space provided.

1. Describe motivational techniques that affect the learning environment.

_____
_____
_____
_____

2. Describe how to adjust the classroom presentation to meet the needs of specific learning objectives.

3. Describe the principles of adult learning.

4. Describe how to deal with disruptive or unsafe behavior by students during instruction.

5. Discuss the characteristics of each generational learner.

Instructor Applications

Discuss your response to these instructor applications to assist you in developing your knowledge of your responsibilities as a fire service instructor.

1. Describe the most effective and least effective training sessions that you have attended, and discuss which methods of instruction were appropriate or inappropriate and how they may have influenced your impression of the training session.

Chapter 3: Methods of Instruction 19

2. Review a list of upcoming training sessions scheduled for your department, and determine the best methods of instruction that should be used to achieve the best outcomes for them.

_____
_____
_____
_____
_____

3. Rearrange your typical classroom seating arrangement to accommodate various methods of discussion such as a lecture, demonstration, or group discussion.

_____
_____
_____
_____
_____

The Learning Process

Workbook Activities

The following activities have been designed to help you. Your instructor may require you to complete some or all of these activities as a regular part of your instructor training program. You are encouraged to complete any activity that your instructor does not assign as a way to enhance your learning in the classroom.

Fire Service Instructor I

Chapter Review

The following exercises provide an opportunity to refresh your knowledge of this chapter.

Multiple Choice

Read each item carefully, and then select the best response.

_____ 1. According to Edward Thorndike, how many laws of learning apply to how behavior is changed through instruction?
   A. 1
   B. 2
   C. 5
   D. 6

_____ 2. Which perspective on learning is reflected when an intellectual process takes place to improve the way you mentally view information?
   A. Psychomotor
   B. Behaviorist
   C. Cognitive
   D. Competency

_____ 3. Which perspective on learning is tied to skills or hands-on training?
   A. Psychomotor
   B. Behaviorist
   C. Cognitive
   D. Competency

_____ 4. According to the laws of learning, which of the following represents the law of recency?
   A. Learning accompanied with satisfaction
   B. The more recent the practice, the more effective the performance
   C. Applying real-life experiences in realistic settings
   D. A person can learn when physically and mentally ready to learn

_____ 5. According to the laws of learning, which of the following represents the law of readiness?
   A. Learning accompanied with satisfaction
   B. The more recent the practice, the more effective the performance
   C. Applying real-life experiences in realistic settings
   D. A person can learn when physically and mentally ready to learn

CHAPTER 4

_____ 6. According to the laws of learning, which of the following represents the law of effect?
   A. Learning accompanied with satisfaction
   B. The more recent the practice, the more effective the performance
   C. Applying real-life experiences in realistic settings
   D. A person can learn when physically and mentally ready to learn

_____ 7. According to the laws of learning, which of the following represents the law of intensity?
   A. Learning accompanied with satisfaction
   B. The more recent the practice, the more effective the performance
   C. Applying real-life experiences in realistic settings
   D. A person can learn when physically and mentally ready to learn

_____ 8. Within the cognitive domain of learning, which of the following is represented by the use of information?
   A. Knowledge
   B. Comprehension
   C. Application
   D. Analysis

_____ 9. Within the cognitive domain of learning, which of the following is represented by the ability to remember information acquired in the past?
   A. Knowledge
   B. Comprehension
   C. Application
   D. Analysis

_____ 10. Within the cognitive domain of learning, which of the following is represented by being able to understand the meaning of information?
   A. Knowledge
   B. Comprehension
   C. Application
   D. Analysis

_____ 11. Which of the following psychomotor domain levels is the earliest that occurs and is defined as simply watching the skill or activity being performed?
   A. Observation
   B. Imitation
   C. Manipulation
   D. Precision

_____ 12. Which of the following psychomotor domain levels is the highest level of performance and is represented by the ability to perform multiple skills correctly?
   A. Manipulation
   B. Precision
   C. Articulation
   D. Naturalization

_____ 13. Which of the domains of learning is often referred to as difficult to measure?
   A. Cognitive
   B. Psychomotor
   C. Affective
   D. Forced learning

_____ 14. If the instructor asks a firefighter to complete a skill using a ground ladder by performing each of the steps required to raise the ladder, he or she is said to be training in which domain?
   A. Cognitive
   B. Psychomotor
   C. Affective
   D. Forced learning

_____ 15. Which theorist developed the Laws of Learning?
   A. Abraham Maslow
   B. Benjamin Bloom
   C. Edward Thorndike
   D. Alan Brunacini

Matching

Match the correct learning domain with its category where learning takes place.

A. Affective   _____ 1. Knowledge
B. Cognitive   _____ 2. Physical use of knowledge
C. Psychomotor   _____ 3. Attitudes, emotions, or values

Match the level of cognitive learning with its proper definition.

D. Knowledge
E. Comprehension
F. Application
G. Analysis
H. Synthesis
I. Evaluation

_____ 4. Integration of information as a whole
_____ 5. Using the information
_____ 6. Remembering knowledge acquired in the past
_____ 7. Understanding the meaning of the information
_____ 8. Breaking the information into parts to help understand all the information
_____ 9. Using standards and criteria to judge the value of the information

Match the classification of learning disability with its short definition.

J. Dysphasia
K. Dyscalculia
L. Dyspraxia
M. Attention-deficit/hyperactivity disorder (ADHD)

_____ 10. A chronic level of inattention and impulsive actions
_____ 11. Difficulty with math subjects
_____ 12. Lack of physical coordination of motor skills
_____ 13. Inability to write, spell, or place words in the right order

True/False

If you believe the statement to be more true than false, write the letter T in the space provided. If you believe the statement to be more false than true, write the letter F.

_____ 1. Learning occurs in either formal or informal settings.
_____ 2. In the psychomotor domain, each level builds on the previous one.
_____ 3. A visual learner as described in the VAK (Visual, Auditory, and Kinesthetic characteristics) model is someone who prefers information to be displayed in words.
_____ 4. There are three levels of affective learning.
_____ 5. The use of teams is not a helpful method of instruction in classroom settings.
_____ 6. A student with ADHD may be protected by the Americans with Disabilities Act.
_____ 7. The acquisition of new values is an example of affective learning.
_____ 8. Knowledge may be defined as remembering past learning.
_____ 9. The term psychomotor refers to the use of the brain and the senses to tell the body what to do.
_____ 10. Using the appropriate teaching style for each learner in a class will not increase safety.

Fill-in

Read each item carefully, and then complete the statement by filling in the missing word(s).

1. _____ is a change in a person's ability to behave in certain ways.

2. Learning takes _____ and _____.

3. Each person has an individual _____ _____, which is the way in which he or she learns the most effectively. This is a key principle for instructors.

4. Bloom's classification system of learning domains is widely known as _____ _____ _____.

5. The law of _____ states that a person can learn when he or she is physically and mentally prepared to respond to instruction.

Short Answer

Complete this section with short written answers using the space provided.

1. Describe the laws and principles of learning.

   _____
   _____
   _____
   _____

2. Identify the three types of learning domains.

   _____
   _____
   _____
   _____

3. Define learning styles, and discuss the effects of learning styles.

   _____
   _____
   _____
   _____

Instructor Applications

Discuss your response to these instructor applications to assist you in developing your knowledge of your responsibilities as a fire service instructor.

1. Review the laws and principles of learning, and identify an instructional method that incorporates each of the laws and principles.

   _____
   _____
   _____
   _____

2. Make a list of potential students you will have in upcoming classes, and based on their knowledge, experience, and other instructional factors, identify the ways each of these student classes learns and how you can make the best applications to them for the class.

_____
_____
_____
_____
_____

3. Review how you would try to increase your students' understanding of basic material by adjusting the lesson plan to their individual needs.

_____
_____
_____
_____
_____

Communication Skills

Workbook Activities

The following activities have been designed to help you. Your instructor may require you to complete some or all of these activities as a regular part of your instructor training program. You are encouraged to complete any activity that your instructor does not assign as a way to enhance your learning in the classroom.

Fire Service Instructor I

Chapter Review

The following exercises provide an opportunity to refresh your knowledge of this chapter.

Multiple Choice

Read each item carefully, and then select the best response.

_____ 1. How many elements are included in a properly functioning communication process?
   A. 2
   B. 3
   C. 4
   D. 5

_____ 2. Which part of the communication process represents the way the message is transmitted between sender and receiver?
   A. Message
   B. Medium
   C. Feedback
   D. Encoding

_____ 3. In the classroom, what does the instructor represent in the communication process?
   A. Sender
   B. Message
   C. Medium
   D. Receiver

_____ 4. An acknowledgment of a radio communication between an incident commander and an officer by repeating back a message represents what part of the communication process?
   A. Message
   B. Medium
   C. Feedback
   D. Affirmation

_____ 5. In the classroom, what does the student represent in the communication process?
   A. Sender
   B. Message
   C. Medium
   D. Receiver

CHAPTER 5

_____ 6. What surrounds the communication process and has a direct influence on learning?
  A. The environment
  B. The message
  C. The medium
  D. The communication style

_____ 7. What should be the format for written communications for outside the organization?
  A. Memo form
  B. SOG form
  C. Letter form
  D. E-mail form

_____ 8. What can hand gestures used by an instructor express?
  A. Aggression
  B. Enthusiasm
  C. Nervousness
  D. All of the above

_____ 9. Your communication style will be:
  A. fixed.
  B. situational.
  C. consistent.
  D. authorative.

_____ 10. Passive listening involves the use of what?
  A. Your eyes
  B. Electronic equipment
  C. Hearing aids
  D. Anticipating the next words to be spoken

_____ 11. A basic rule of thumb for writing is to use the five:
  A. W's.
  B. H's.
  C. A's.
  D. C's.

_____ 12. Which of the following is considered an "informal" form of written communication?
  A. Letter to a citizen
  B. Department operating procedure
  C. Memo regarding a shirt order
  D. Monthly activity report

_____ 13. Which type of presentation would be most effective when teaching fire behavior to rookie fire fighters?
  A. Case studies
  B. Role play
  C. Group discussion
  D. Lecture

____ 14. Which part of the communication process is considered the most "complex"?
   A. Sender
   B. Message
   C. Feedback
   D. Medium

____ 15. Noise created by nearby traffic and heard while at a training tower is considered:
   A. nonverbal noise.
   B. verbal noise.
   C. ambient noise.
   D. disruptive noise.

Matching

Match the element of the communications process with its simple definition.

A. Sender
B. Message
C. Medium
D. Receiver
E. Feedback

____ 1. The student
____ 2. An acknowledgment or action that completes the communication process
____ 3. How the message is transmitted
____ 4. The initiator of the process
____ 5. What is being communicated

True/False

If you believe the statement to be more true than false, write the letter T in the space provided. If you believe the statement to be more false than true, write the letter F.

____ 1. For the communications chain to function properly, all of the links need to be attached to one another.
____ 2. One of the best ways to develop and improve your communication skills is to practice them.
____ 3. After you have mastered the communication process, the learning environment becomes less of a factor in the learning process.
____ 4. People who talk are always good communicators.
____ 5. Speaking softly may be an effective tool in communications.
____ 6. A student's reading level should not influence the instructor's delivery of information or the rate of learning.
____ 7. Group exercises and role playing can be effective methods of communication and instruction, especially with experienced students.
____ 8. The basic rule for writing is to include the three W's and the H.
____ 9. By simply raising or lowering your voice, you can emphasize important points.
____ 10. The fire service has its own unique language.

Fill-in

Read each item carefully, and then complete the statement by filling in the missing word(s).

1. In many learning situations, _____ breeds _____ in the learning process.

2. Communication may be _____ percent what you say and _____ percent how you say it.

3. The two types of listening are _____ and _____.

4. In order to be an effective instructor, you must become an effective _____.

5. Most presentations for fire service instructors involve _____ _____ or _____.

6. Most of the writing you do as a fire fighter and as a fire service instructor will be _____ _____ _____.

Short Answer
Complete this section with short written answers using the space provided.

1. Describe the elements of the communication process.
   _____

2. Describe the role of communication in the learning process.
   _____

3. Compare the different types of communication.
   _____

Instructor Applications
Discuss your response to these instructor applications to assist you in developing your knowledge of your responsibilities as a fire service instructor.

1. List the common distracters you have witnessed when being taught by your instructors. Describe how they impaired the learning process for you.
   _____
   _____
   _____
   _____
   _____

2. Describe the traits of a person who you felt was a great communicator in the classroom.
   _____
   _____
   _____
   _____
   _____

3. List the types of news stories or political speeches that have held your interest, and determine why they did so.
   _____
   _____
   _____
   _____
   _____

Lesson Plans

Workbook Activities

The following activities have been designed to help you. Your instructor may require you to complete some or all of these activities as a regular part of your instructor training program. You are encouraged to complete any activity that your instructor does not assign as a way to enhance your learning in the classroom.

Fire Service Instructor I

Chapter Review

The following exercises provide an opportunity to refresh your knowledge of this chapter.

Multiple Choice

Read each item carefully, and then select the best response.

_____ 1. What is typically the starting point for lesson plan development that the content of the lesson plan is driven by?
   A. Learning objectives
   B. Preparation step
   C. Application step
   D. Evaluation instruments

_____ 2. One of the primary reasons to use a lesson plan is to:
   A. provide direction for class outcome.
   B. organize all material in a logical order.
   C. identify the evaluation process.
   D. All of the above

_____ 3. In the ABCD method of objective development, what does the C stand for?
   A. Consistent
   B. Concise
   C. Confident
   D. Condition

_____ 4. How many components should be present in a properly prepared learning objective?
   A. 1
   B. 2
   C. 3
   D. 4

_____ 5. What does the behavior of the learning objective describe?
   A. The student
   B. The situation in which the student will perform
   C. An observable and definable action
   D. How well the action is performed

CHAPTER 6

_____ 6. What does the degree of the learning objective describe?
   A. The student
   B. The situation in which the student will perform
   C. An observable and definable action
   D. How well the action is performed

_____ 7. What does the audience of the learning objective describe?
   A. The student
   B. The situation in which the student will perform
   C. An observable and definable action
   D. How well the action is performed

_____ 8. What does the condition of the learning objective describe?
   A. The student
   B. The situation in which the student will perform
   C. An observable and definable action
   D. How well the action is performed

_____ 9. Which component of the lesson plan contains the main body or the outline of the material to be presented?
   A. Objective statement
   B. Lesson outline
   C. Resources and references
   D. Summary

_____ 10. In the four-step method of instruction, which step motivates and gets the student ready to learn?
   A. Preparation
   B. Presentation
   C. Application
   D. Evaluation

_____ 11. In the four-step method of instruction, which step gives the student the ability to demonstrate his or her comprehension of lesson material?
   A. Preparation
   B. Presentation
   C. Application
   D. Evaluation

_____ 12. Which step in the four-step method of instruction is considered the most important step?
   A. Preparation
   B. Presentation
   C. Application
   D. Evaluation

_____ 13. In terms of lesson plan development, which of the following statements represents the duties of the Instructor I in terms of adjusting a lesson plan from its original state?
   A. An Instructor I will alter the lesson plan.
   B. An Instructor I will modify the lesson plan.
   C. An Instructor I will create the lesson plan.
   D. An Instructor I cannot make any adjustments to a lesson plan.

_____ 14. Information regarding the prerequisites for a course will be identified in which section of the lesson?
   A. Level of instruction
   B. Reference
   C. Lesson outline
   D. Preparation

_____ 15. If a lesson plan requires the use of a self-contained breathing apparatus (SCBA) unit as a visual aid, where would this information be located in the lesson plan?
   A. Learning objectives
   B. Materials needed
   C. References
   D. Application step

Matching

Match the part of the lesson plan with its simple definition.

A. Lesson title
B. Level of instruction
C. Behavioral objectives
D. Instructional materials
E. Lesson outline
F. References/resources
G. Lesson summary
H. Assignment

_____ 1. The review of the lesson plan
_____ 2. Homework or reading assignments
_____ 3. The description of what the lesson is about
_____ 4. Listing of props and instructional aids
_____ 5. The main body of the lesson plan
_____ 6. Identification of sources from which lesson content was derived
_____ 7. The specific outcomes of the lesson plan
_____ 8. Identification of the students who should be in the class

Match the name of each step of the four-step method of instruction with its simple definition.

A. Preparation
B. Presentation
C. Application
D. Evaluation

_____ 9. The students' use of the knowledge presented in the class
_____ 10. The body and actual delivery of the lesson plan
_____ 11. The motivational step that gets the student ready to learn
_____ 12. The measure of students' knowledge or skills

Chapter 6: Lesson Plans

True/False
If you believe the statement to be more true than false, write the letter T in the space provided. If you believe the statement to be more false than true, write the letter F.

_____ 1. It is not necessary to present all elements of a behavioral objective in order.

_____ 2. The Fire Service Instructor I cannot modify a lesson plan objective that identifies the number of fire fighters who will perform an evolution.

_____ 3. The Fire Service Instructor I cannot modify the lesson plan and course materials to meet the needs of the students and their learning styles.

_____ 4. A lesson plan should be reviewed to make sure it meets local standard operating procedures (SOPs).

_____ 5. Grading criteria are not part of the lesson plan but are based on the assignment for the class.

_____ 6. There is only one format for lesson plans used in the fire service—the ABCD format.

_____ 7. Lesson plans are like maps that help guide the course of a learning process.

_____ 8. The terminal objective is the destination to which the instructor is leading the student.

_____ 9. The evaluation step of the four-step method of instruction ensures that students correctly acquired knowledge and skills.

_____ 10. During the presentation step of the instructional process, the instructor prepares the student to learn through various motivational techniques.

Fill-in
Read each item carefully, and then complete the statement by filling in the missing word(s).

1. The Fire Service Instructor I should not _____ the content or the _____ of a lesson plan.

2. The Fire Service _____ presents the content of a lesson plan.

3. A(n) _____ is the goal that is achieved through attainment of a knowledge or skill.

4. You must identify the _____ in a lesson plan because the student must be able to understand the material.

5. The Instructor I may _____ a lesson plan to fit students' needs, whereas the Instructor II may _____ the lesson plan to change the fundamental content of it.

Short Answer
Complete this section with short written answers using the space provided.

1. List the components of learning objectives.

   _____

2. List and describe the parts of a lesson plan.

   _____

3. List the four-step method of instruction.

   _____

4. Describe the difference between "adapting" a lesson plan versus "modifying" a lesson plan.

   _____

Instructor Applications

Discuss your response to these instructor applications to assist you in developing your knowledge of your responsibilities as a fire service instructor.

1. Using a sample lesson plan provided in class, identify each component.
   _____

2. Using the same lesson plan, identify what parts of the lesson plan would be adjusted to accommodate special needs of students.
   _____

3. Identify a locally developed lesson plan and determine whether all of the required elements of it are present.
   _____

Fire Service Instructor II

Multiple Choice
Read each item carefully, and then select the best response.

_____ 1. What are the three lowest levels of cognitive objectives?
   A. Evaluation, synthesis, application
   B. Knowledge, comprehension, application
   C. Condition, behavior, standard
   D. Preparation, presentation, application

_____ 2. Which of the following is the main goal of developing a lesson plan?
   A. Meeting the needs of your agency
   B. Creating a document that can be used by any qualified instructor
   C. Creating a document that will ensure that students achieve the learning objective
   D. All of the above

_____ 3. A simple way of developing an outline for your lesson is to use which of the following formats?
   A. ABCD
   B. ADDIE
   C. Introduction, body, conclusion
   D. Objective, content, evaluation

_____ 4. What is the first step of lesson plan development?
   A. Determine the learning objective.
   B. Determine the course goal.
   C. Determine the audience for the lesson.
   D. Determine the training need for the department.

_____ 5. Which part of the learning objectives identifies *who* the lesson is being developed for?
   A. Behavior
   B. Audience
   C. Lesson summary
   D. Lesson outline

_____ 6. Which part of the learning objectives best describes the measurable action element?
   A. Audience
   B. Condition
   C. Degree
   D. Behavior

_____ 7. Who developed the Taxonomy of Education Objectives, which is used in the development of lesson plans?
   A. Heinrich
   B. Russell
   C. Bloom
   D. Reeder

_____ 8. Which of the following are objectives found in the cognitive domain?
   A. Knowledge, comprehension, application
   B. Imitation, manipulation, articulation
   C. Responding, valuing, organizing
   D. Knowledge, manipulation, valuing

_____ 9. Which of the following are objectives found in the psychomotor domain?
   A. Knowledge, comprehension, application
   B. Imitation, manipulation, articulation
   C. Responding, valuing, organizing
   D. Knowledge, manipulation, valuing

_____ 10. Which of the following are objectives found in the affective domain?
   A. Knowledge, comprehension, application
   B. Imitation, manipulation, articulation
   C. Responding, valuing, organizing
   D. Knowledge, manipulation, valuing

_____ 11. Using the job performance requirement terminology, the word "given" best describes which part of the learning objective?
   A. Audience
   B. Behavior
   C. Attitude
   D. Condition

_____ 12. When developing a lesson plan, which part describes the student's success in meeting the goals of the lesson?
   A. Degree
   B. Condition
   C. Assignment
   D. Performance

_____ 13. Which organization has developed the job performance requirements for use when developing lesson plans?
   A. OSHA
   B. ISFSI
   C. NFPA
   D. IAFC

_____ 14. Which component of the lesson plan development process is considered the "main" body?
   A. Lesson objectives
   B. Lesson outline
   C. Lesson summary
   D. Lesson outcomes assessment

_____ 15. Which of the following is an important final step after you have modified a lesson plan?
   A. Notify the chief that you have modified the lesson plan.
   B. Update the references in the lesson plan.
   C. Evaluate the lesson plan to determine if it meets the lesson objectives.
   D. Retain a copy of the original lesson plan for future reference if necessary.

Matching

Match each of the items in the left column with the appropriate description in the right column.

_____ 1. Audience
_____ 2. Behavior
_____ 3. Condition
_____ 4. Degree
_____ 5. Cognitive domain
_____ 6. Psychomotor domain
_____ 7. Affective domain
_____ 8. Job performance requirements
_____ 9. Psychomotor lesson plans
_____ 10. Behavior

A. Includes demonstration and displaying
B. What the students will do or learn
C. Must take safety concerns into consideration
D. The situation in which the student will perform the behavior
E. Is defined by such verbs as challenge and attempt
F. Includes analysis, evaluation, and comprehension
G. Determining this requires measureable action words
H. The *who* element of a lesson plan
I. The extent to which students have learned
J. Are defined by the NFPA

True/False

If you believe the statement to be more true than false, write the letter T in the space provided. If you believe the statement to be more false than true, write the letter F.

_____ 1. The Fire Service Instructor II is responsible for creating lesson plans.
_____ 2. A job performance requirement is converted into a learning objective when developing a lesson plan.
_____ 3. A knowledge-based objective requires the student to have the ability to solve problems or apply information.
_____ 4. Only a Fire Service Instructor II will present a prepared lesson plan.
_____ 5. Using Karthwohl's revision, "knowledge" and "understanding" are the same thing.
_____ 6. A Fire Service Instructor II determines which domain a lesson plan addresses.
_____ 7. Accept and participate are action verbs found in the affective domain.
_____ 8. Practice and assemble are action verbs found in the affective domain.
_____ 9. The most common format for a lesson plan is the two-column format.
_____ 10. Often the inclusion of one instruction aid creates a need for more instructional materials.

Fill-In

Read each item carefully, and then complete the statement by filling in the missing word(s).

1. Using a(n) _____ to develop a lesson plan adds _____ in training.

2. Whenever tasked with developing a lesson plan for a class, start by _____ the intended outcome with the person requesting the class.

3. The _____ is the main body of the lesson plan.

4. Each part of the _____ plan should be directly tied to the _____ objectives.

5. Utilization of fire service reference and NFPA job performance requirements provides content _____ to the material being taught.

Short Answer

Complete this section with a short written answer using the space provided.

1. List the three domains used when developing behavioral objectives for a lesson plan.
   _____

Instructor Applications

Discuss your response to these instructor applications to assist you in developing knowledge about your responsibilities as a fire service instructor.

1. Review a lesson plan and determine what components of the lesson plan would have to be modified if you were to change one of the learning objectives.
   _____

2. Review a lesson plan and identify the different domains by the appropriate action verb.
   _____

3. Identify your agency's policy for modifying lesson plans.
   _____

The Learning Environment

Workbook Activities

The following activities have been designed to help you. Your instructor may require you to complete some or all of these activities as a regular part of your instructor training program. You are encouraged to complete any activity that your instructor does not assign as a way to enhance your learning in the classroom.

Fire Service Instructor I, II

Chapter Review

The following exercises provide an opportunity to refresh your knowledge of this chapter.

Multiple Choice

Read each item carefully, and then select the best response.

_____ 1. The makeup of your audience based on race, gender, economics, or ethnic origin is referred to as:
   A. demographics.
   B. background information.
   C. diversity.
   D. democracy.

_____ 2. In general, there are _____ types of fire departments?
   A. two
   B. three
   C. five
   D. seven

_____ 3. If students in a class that you are preparing to teach are not at the level of knowledge or understanding required for the material you are to present, you should:
   A. not teach the lesson plan as assigned and instead teach something else.
   B. call the lead instructor and ask them what to do.
   C. modify the lesson plan to meet the students' needs but teach the lesson plan as assigned.
   D. rewrite the lesson plan and change the lesson objectives.

_____ 4. Which of the following is a disadvantage to a hollow square seating arrangement?
   A. Students can see each other.
   B. It is an ideal setup for small-group discussion.
   C. It allows students the opportunity to develop ideas.
   D. Students can easily be distracted or have off-subject conversations.

_____ 5. The best seating arrangement for conducting a demonstration of donning an SCBA would be a:
   A. large U-shaped arrangement.
   B. hollow square.
   C. small V-shaped arrangement.
   D. T arrangement.

CHAPTER 7

_____ 6. One of the more often overlooked areas of a learning environment is:
   A. lighting.
   B. room temperature.
   C. noise distractions.
   D. All of the above

_____ 7. Which of the following statements is NOT correct?
   A. Eliminating noise distractions outside is very difficult.
   B. Informal training takes place in every environment.
   C. Small-group activities have limited use in the fire service.
   D. Classroom lighting is an important part of the learning process.

_____ 8. The greatest concern in the outdoor classroom is:
   A. noise.
   B. weather.
   C. safety.
   D. distractions.

_____ 9. When teaching during a live fire evolution, you should do which of the following to communicate?
   A. Remove your face piece so students can hear you.
   B. Use hand signals to instruct the session.
   C. Use a calm and controlled voice to communicate the objective you are teaching.
   D. All of the above

_____ 10. Prerequisites are used for all of the following EXCEPT:
   A. determining successful completion of a course.
   B. ensuring that students entering a course are at a certain level of knowledge.
   C. satisfying NFPA requirements.
   D. limiting attendance to only students who pass an entrance exam.

_____ 11. The highest level of student success in the classroom environment, as adapted from Maslow's hierarchy of needs, is:
   A. a warm learning environment.
   B. the student successfully completing the course assignment.
   C. the student having a feeling of safety in the learning environment.
   D. the student being allowed to fail without worry of punishment.

_____ 12. (1) Asking students to turn off cell phones before beginning instruction is not appropriate in today's world. (2) Limiting student access to the class until a break in the lecture is an acceptable method for controlling the classroom environment.
   A. Both statements are correct.
   B. Both statements are incorrect.
   C. Only statement 1 is correct.
   D. Only statement 2 is correct.

_____ 13. The basic or first level of learning, as adapted from Maslow's hierarchy of needs, is:
   A. a warm learning environment.
   B. the student successfully completing the course assignment.
   C. the student having a feeling of safety in the learning environment.
   D. the student being allowed to fail without worry of punishment.

_____ 14. What is the biggest difference between a training session that is conducted indoors versus outdoors?
  A. Lighting
  B. Noise
  C. Weather
  D. Setup needs

_____ 15. Every fire department has its own unofficial way of doing business, which is referred to as a department's:
  A. makeup.
  B. culture.
  C. policy.
  D. history.

Matching

Match each of the items in the left column with the appropriate description in the right column.

_____ A. Learning environment
_____ B. Demographics
_____ C. Lecture
_____ D. Discussion
_____ E. Large U shape
_____ F. Prerequisite
_____ G. Classroom environment
_____ H. NFPA 1403
_____ I. Evaluation
_____ J. Contingency plan

1. Standard on Live Fire Training
2. May occur in auditorium, traditional classroom, or theater
3. Factors that influence a person's behavior based on gender, age, or marital status
4. Used to demonstrate skills for a class
5. The perfect environment that satisfies all of Maslow's hierarchy of needs
6. Plan B
7. Environment where ideas are shared and learning is achieved
8. Often occurs at a conference table or in a hollow square
9. Tool used to determine success of a student's increase in knowledge
10. Tool used to determine a student's starting point

True/False

If you believe the statement to be more true than false, write the letter T in the space provided. If you believe the statement to be more false than true, write the letter F.

_____ 1. Instructor level I is responsible for developing lesson plans.
_____ 2. Instructor level II determines instructor-to-student ratio.
_____ 3. Age is considered part of demographic information.
_____ 4. Offensive language in the classroom is normal and acceptable because of the nature of the fire service.
_____ 5. Typically, students are not concerned with course objectives.
_____ 6. It is acceptable to modify course objectives based on information obtained during the introduction of the class.
_____ 7. Content of the course material is more important than the classroom setup.
_____ 8. Breaks during classroom instruction should be limited to one every 2 hours.
_____ 9. There are no safety issues during classroom-only instruction.
_____ 10. Some of the greatest lessons learned from the fire ground take place on a tailboard in a truck bay after an incident.

Fill-in

Read each item carefully, and then complete the statement by filling in the missing word(s).

1. Department _____ will influence how you teach and will require you to adapt your training accordingly.

2. A _____ training center with tables and chairs puts students in the mind frame that students' _____ should match the location where they are.

3. A well-prepared instructor will always have a(n) _____ plan in the event of bad weather.

4. Having the right number of _____, in the right position, doing the right things, with an emphasis on _____ makes the learning environment complete.

5. If you hear an off-color joke and _____, you are _____.

Short Answer
Complete this section with short written answers using the space provided.

1. Describe what encompasses the word "demographics."

2. Describe what should be considered in creating a learning environment.

3. Describe how Maslow's hierarchy of needs can be adapted to a learning environment.

Instructor Applications
Discuss your response to these instructor applications to assist you in developing your knowledge of your responsibilities as a fire service instructor.

1. Review an upcoming class you are scheduled to teach, and look at the demographics of the audience to see whether it changes how you prepare for this course.

2. Review your department's policy on instructor behavior or conduct. If your department does not have one, suggest the development of one and offer to help in its development.

3. Review your department's policy on classroom safety and training ground safety. Determine when it was last updated, and determine how encompassing it is.

Technology in Training

Workbook Activities

The following activities have been designed to help you. Your instructor may require you to complete some or all of these activities as a regular part of your instructor training program. You are encouraged to complete any activity that your instructor does not assign as a way to enhance your learning in the classroom.

Fire Service Instructor I, II

Chapter Review

The following exercises provide an opportunity to refresh your knowledge of this chapter.

Multiple Choice

Read each item carefully, and then select the best response.

_____ 1. What is a key element of most presentations?
   A. Multimedia tools
   B. Learning environment
   C. Clear speaker
   D. Room arrangement

_____ 2. Which of the following statements is correct concerning today's new fire fighters?
   A. New fire fighters have not been exposed to modern technology.
   B. New fire fighters were raised using technology in the learning environment.
   C. New fire fighters like lectures and simulations.
   D. New fire fighters lack technology skills.

_____ 3. When using a PowerPoint presentation, which of the following should you NOT do?
   A. Avoid reading from the slides.
   B. Avoid standing in front of the screen.
   C. Turn the lights off so everyone can read the slides.
   D. Have a backup plan in case the system does not work.

_____ 4. When developing a PowerPoint slide, what is the ideal number of lines and words per line?
   A. 5 lines, 7 words per line
   B. 7 lines, 7 words per line
   C. 7 lines, 5 words per line
   D. 6 lines, 6 words per line

_____ 5. Which type of projector is most commonly used to present PowerPoint presentations and videos in a large classroom?
   A. VHS player
   B. ACD projector
   C. LCD projector
   D. DVD player

CHAPTER 8

_____ 6. Which of the following is an advantage of a PowerPoint presentation?
   A. Embedding video and audio into your presentation
   B. Reading from the slide if you forget your lecture notes
   C. Overloading the students' senses with information so they will not ask questions
   D. Using technology to increase the learning curve

_____ 7. Which of the following is NOT an advantage of videotapes?
   A. They are inexpensive.
   B. There is a wide selection available for use.
   C. They are copyrighted material.
   D. They can be viewed on any television with a VCR.

_____ 8. When purchasing a digital camera, you should buy one with a minimum of _____ megapixels.
   A. 10
   B. 21
   C. 8
   D. 4

_____ 9. Which of the following is an online learning platform?
   A. Angel
   B. MODEL
   C. Whiteboard
   D. Webbing GT

_____ 10. Which of the following devices has been used for over 40 years?
   A. Computer
   B. Overhead projector
   C. Slide projector
   D. LCD projector

_____ 11. Which of the following is an example of a hand-held personal computer?
   A. ACB
   B. iPod
   C. PDA
   D. LCD

_____ 12. When creating PowerPoint slides, a rule is that headings should be _____-point font, and the body text should be _____-point font.
   A. 42, 32
   B. 56, 38
   C. 28, 36
   D. 38, 28

_____ 13. An important part of modern maintenance is having what?
   A. Current software for the computer or device
   B. Extra light bulbs
   C. Cleaning solutions
   D. Router cleaner

_____ 14. Which of the following is NOT a reason for using multimedia for presentations?
    A. Multimedia presentations allow students to see, hear, and in some cases touch the learning process.
    B. Multimedia entertains students and takes pressure off the instructor.
    C. Multimedia simplifies complex theories into an understandable format.
    D. Multimedia keeps today's students interested in learning.

_____ 15. Which of the following is the biggest issue facing instructors who want to use multimedia?
    A. The cost
    B. The students' learning level
    C. Determining when it is appropriate
    D. The instructors' unwillingness to learn about new technology

Matching

Match the following terms with the correct description.

A. Hybrid training
B. Visual projector
C. Adobe Flash Player
D. Fire and Emergency Training Network
E. Audio sound system
F. Blackboard (www.blackboard.com)
G. PowerPoint presentation
H. National Emergency Training Center
I. Smart board
J. Model

_____ 1. Abbreviated PPT
_____ 2. Includes wireless microphone, transmitter, and receiver
_____ 3. An interactive whiteboard that uses touch detection for user input.
_____ 4. Projects a page from a textbook or a map or diagram
_____ 5. Uses file extension "swf"
_____ 6. Provides computer-based training
_____ 7. Combines classroom and Internet learning
_____ 8. Example of a virtual academy
_____ 9. Provides satellite programming
_____ 10. Type of learning that occurs in a controlled environment

True/False

If you believe the statement to be more true than false, write the letter T in the space provided. If you believe the statement to be more false than true, write the letter F.

_____ 1. Handouts are considered a form of media.
_____ 2. Chalkboards are considered "healthier" than erasable boards.
_____ 3. An advantage of easel pads is the ability to save the information and post it around a classroom.
_____ 4. Clip art files provide pictures from newspapers that can be copied and passed out to a class.
_____ 5. Quick-time media programs are used to develop PowerPoint slides.
_____ 6. PowerPoint slides should support a lecture or presentation.

_____ 7. Ghosting is a term that refers to faint shadows that appear to the right of letters or numbers on a slide.

_____ 8. WebCT is an example of software used to embed video into a PowerPoint presentation.

_____ 9. The proper use of multimedia methodology in the learning process will increase the student's ability to understand the information.

_____ 10. When using multimedia during a presentation, an instructor should have contingency plans A, B, and C.

Fill-in
Read each item carefully, and then complete the statement by filling in the missing word(s).

1. Multimedia tools come in a(n) _____ of types and are intended for multiple uses.

2. A disadvantage of multimedia is the potential for _____-_____ on technology as part of your instruction.

3. An advantage of a(n) _____ is the ability to load printed material onto your computer.

4. A(n) ___ ___ ___ allows students to download lesson material and listen to it on an iPod.

5. A(n) _____ can be used to teach incident command in a controlled classroom environment.

Short Answer
Complete this section with short written answers using the space provided.

1. What can you do as an instructor to learn to use media for your presentations?
   _____

2. What are three methods for troubleshooting projected media?
   _____

3. What is an advantage of using multimedia during a presentation?
   _____

Instructor Applications
Discuss your response to these instructor applications to assist you in developing your knowledge of your responsibilities as a fire service instructor.

1. Experiment with a PowerPoint presentation. Which color schemes work best for dark classrooms? Which work best for bright classrooms? Learn how to change a PowerPoint presentation quickly so that you can instantly adapt to existing conditions.
   _____

2. Modify a PowerPoint presentation by adding text, photos, or other illustrations to clarify the message.
   _____

3. Select a learning platform such as an online or distance learning course. Discover what the benefits of that platform are for your students.
   _____

Safety During the Learning Process

Workbook Activities

The following activities have been designed to help you. Your instructor may require you to complete some or all of these activities as a regular part of your instructor training program. You are encouraged to complete any activity that your instructor does not assign as a way to enhance your learning in the classroom.

Fire Service Instructor I, II

Chapter Review

The following exercises provide an opportunity to refresh your knowledge of this chapter.

Multiple Choice

Read each item carefully, and then select the best response.

_____ 1. As the instructor in a training situation, you should:
  A. always put safety ahead of course objectives.
  B. always put course objectives ahead of outcome assessments.
  C. always keep cost under budget to ensure continued funding.
  D. always put instructor safety ahead of trainee safety.

_____ 2. One of the 16 Fire Fighter Life Safety Initiatives is to reduce line of duty deaths by:
  A. 25 percent over the next 10 years.
  B. 50 percent over the next 10 years.
  C. 70 percent over the next 5 years.
  D. 50 percent over the next 5 years.

_____ 3. When it comes to safety, fire service instructors should:
  A. delegate safety to others.
  B. follow NFPA 1404 during live fire training.
  C. lead by example in everything they do.
  D. expect others to do as they say, not as they do.

_____ 4. Safety during training begins:
  A. with the attitude of the fire chief.
  B. in development of training objectives.
  C. once you hit the training ground.
  D. with the classroom environment.

_____ 5. An important rule is that the level of Personal Protective Equipment required for an emergency is:
  A. the level of PPE that should be worn during training.
  B. a minimum standard set by the NFPA.
  C. not required during training.
  D. determined by each incident.

_____ 6. Which NFPA standard establishes the policy for rehabilitation during training?
   A. NFPA 1403
   B. NFPA 1584
   C. NFPA 1500
   D. NFPA 1710

_____ 7. Failure by the instructor to correct improper skill performance on the training ground can lead to:
   A. the fire fighter passing the course material.
   B. the company officer sending a letter to the training chief.
   C. unsafe performance during a real-life incident.
   D. failing the next evaluation.

_____ 8. Which of the following would NOT be considered high-risk training?
   A. Swift water training
   B. Live fire evolutions
   C. Liquefied petroleum gas fire evolutions
   D. Trench simulator training

_____ 9. Which of the following Incident Command System positions is mandated to be staffed during a live fire evolution?
   A. Safety officer
   B. Water supply officer
   C. Rehab officer
   D. Operations officer

_____ 10. According to NFPA 1403, the person responsible for planning and coordinating all training activities is the:
   A. incident commander.
   B. instructor in charge.
   C. safety officer.
   D. lead instructor.

_____ 11. When selecting instructors to assist in the training process, you should consider all of the following EXCEPT:
   A. their attitude toward safety.
   B. their skills and qualifications.
   C. their gender or race.
   D. their presentation skills.

_____ 12. (1) Tragedies during training are usually single events that occur without warning. (2) Accidents or injuries that occur during training are usually a series of small problems that build into an event or accident.
   A. Only statement 1 is correct.
   B. Only statement 2 is correct.
   C. Both statements are correct.
   D. Both statements are incorrect.

____ 13. During an accident investigation, it is important to:
   A. control information released by the department.
   B. be selective in who is interviewed.
   C. seek answers, not blame.
   D. determine who is at fault.

____ 14. (1) Students have a responsibility for their own safety. (2) Students have a responsibility for meeting all of a course's prerequisites.
   A. Only statement 1 is correct.
   B. Only statement 2 is correct.
   C. Both statements are correct.
   D. Both statements are incorrect.

____ 15. Who is responsible for staying current on laws, regulations, and standards that affect fire fighter training?
   A. Fire chief
   B. Safety officer
   C. City attorney
   D. Training officer

Matching

Match each of the terms in the left column to the appropriate definition in the right column.

A. Fire Service Instructor I responsibilities
B. Five to one
C. Hands-on training
D. Department training SOPs
E. Fire Instructor II responsibilities
F. Sixteen Life-Safety Initiatives
G. Classroom safety
H. Low-risk training
I. Initiative number 5
J. Hidden hazards during training process

____ 1. An effort aimed at reducing fire fighter line of duty deaths
____ 2. Train to national standards applicable to all fire fighters based on duties assigned
____ 3. Involves identifying exits and procedures in the event of an emergency
____ 4. Involves setting up ladders and climbing to a roof
____ 5. Use of incident command, rehabilitation, and use of full PPE during training
____ 6. Supervising other instructors and using safety policies
____ 7. Requires instructors to provide a learning environment that protects the emotional side of the fire fighter
____ 8. Student-to-instructor ratio during high-risk training
____ 9. Following lesson outline and adhering to department safety policies
____ 10. Practicing knots in a classroom environment

True/False

If you believe the statement to be more true than false, write the letter T in the space provided. If you believe the statement to be more false than true, write the letter F.

____ 1. On average, 10 fire fighters die each year during training activities.
____ 2. The Sixteen Fire Fighter Life-Safety Initiatives are mandatory requirements.
____ 3. It is acceptable to modify PPE requirements during training.
____ 4. One of the main causes of injuries during training is the failure to follow SOPs.

_____ 5. The strongest message you send as an instructor is what you say in the classroom.
_____ 6. Safety in the training environment should be stated in the lesson plan.
_____ 7. The use of an unsafe chain saw is better than not using one at all.
_____ 8. It is better to stress or push a new recruit during training than to let him or her fail on the fire ground.
_____ 9. All prerequisite training must be met prior to participating in a live fire exercise.
_____ 10. During live fire training, the instructor in charge fills the role of the Incident Commander.

Fill-in

Read each item carefully, and then complete the statement by filling in the missing word(s).

1. All fire fighters must be _____ to stop unsafe practices.

2. Use available _____ wherever it can produce higher levels of health and _____.

3. The fire service has been _____ in developing methods and practices to accomplish drills and training.

4. An adequate number of _____ must be present to conduct training safely.

5. Fire-ground _____ translate to the _____ _____ , and you should apply your fire fighter experience to every evolution you instruct.

Short Answer

Complete this section with short written answers using the space provided.

1. How can you adapt the 16 Fire Life-Safety Initiatives into your department's training activities?
   _____

2. How can you teach safety to a profession that by its nature operates in a dangerous environment?
   _____

3. How do you develop a safety culture?
   _____

Instructor Applications

Discuss your response to these instructor applications to assist you in developing your knowledge of your responsibilities as a fire service instructor.

1. Review local policies relating to safety in training.
   _____

2. Read NFPA 1403, Standard on Live Fire Training, and review the sample checklists used for conducting live fire training exercises.
   _____

3. Research fire fighter line-of-duty deaths in training accidents, and review the recommendations of the investigators regarding ways to prevent similar occurrences.
   _____

Evaluating the Learning Process

Workbook Activities

The following activities have been designed to help you. Your instructor may require you to complete some or all of these activities as a regular part of your instructor training program. You are encouraged to complete any activity that your instructor does not assign as a way to enhance your learning in the classroom.

Fire Service Instructor I, II

Chapter Review

The following exercises provide an opportunity to refresh your knowledge of this chapter.

Multiple Choice

Read each item carefully, and then select the best response.

_____ 1. All of the following are reasons for testing **EXCEPT**:
   A. to justify the purchase of new equipment.
   B. to measure student attainment of learning objectives.
   C. to enhance and improve training programs.
   D. to identify and determine weaknesses and gaps in training programs.

_____ 2. Which of the following testing tools is **not** used in the fire service very often?
   A. Written
   B. Oral
   C. Performance
   D. Psychomotor

_____ 3. A test that measures the students' ability to perform a task is an example of a(n):
   A. oral test.
   B. written test.
   C. skills evaluation.
   D. cognitive test.

_____ 4. Who must give permission for the release of students' test scores?
   A. The instructor
   B. The student
   C. The chief
   D. The training officer

CHAPTER 10

_____ 5. What is the most correct method for releasing a student's test score?
   A. Post scores with social security numbers on the bulletin board.
   B. Read off the names and scores in the classroom during class time.
   C. One-on-one basis in a private office.
   D. You cannot release student test scores but must send them to the department for release.

_____ 6. When proctoring a written exam, you should do all of the following **EXCEPT**:
   A. arrive early and set up the classroom.
   B. space students away from each other as much as possible.
   C. monitor students at all times, never leaving the testing room.
   D. give out answers to questions that you feel are incorrect.

_____ 7. If you observe a student cheating during a written exam, you should:
   A. document the activity but let the student continue taking the test.
   B. ask the student to leave and tear up the test.
   C. follow department policy that addresses this issue.
   D. challenge the student during the test and ask him or her to leave.

Matching

Match each of the terms in the left column to the appropriate definition in the right column.

A. Oral test
B. Short-answer essay
C. Performance evaluation

_____ 1. Written testing tool
_____ 2. "Given a hand tool, explain in detail how to clean the tool"
_____ 3. Don an SCBA within 1 minute

True/False

If you believe the statement to be more true than false, write the letter T in the space provided. If you believe the statement to be more false than true, write the letter F.

_____ 1. Proctoring a written exam is the same as evaluating a skills exam.
_____ 2. As an instructor, you should never stop a skills exam or evaluation.
_____ 3. Providing feedback to students is just as important as teaching a lesson.

Fill-in

Read each item carefully, and then complete the statement by filling in the missing word(s)

1. A sound _____ program allows you to know whether a student is progressing in the learning process.

2. A _____ _____ can be made up of eight types of test items.

3. A _____ will address what an instructor should do in the event of a student cheating on an exam.

Short Answer

Complete this section with short written answers using the space provided.

1. Describe some of the common problems with testing.
   _____
   _____

2. Describe the four forms of test-item validity.
   _____
   _____

Fire Service Instructor II

Chapter Review

The following exercises provide an opportunity to refresh your knowledge of this chapter.

Multiple Choice

Read each item carefully, and then select the best response.

_____ 1. Which of the following terms is used to describe how well a test item measures what the developer intended it to measure?
   A. Reliability
   B. Currency
   C. Dependability
   D. Validity

_____ 2. Which of the following is considered the lowest form of test validity?
   A. Face
   B. Technical content
   C. Job content
   D. Currency

_____ 3. _____ means the test item measures the knowledge in a consistent manner.
   A. Reliability
   B. Currency
   C. Dependability
   D. Validity

_____ 4. Which written test is the most widely used in the fire service?
   A. True/false
   B. Multiple choice
   C. Matching
   D. Fill in the blank

_____ 5. An important advantage of the multiple choice test item is that it can be used:
   A. to weight a test.
   B. to distract the unlearned student from the correct answer.
   C. to measure higher mental functions such as reasoning and judgment.
   D. to ask valid and invalid questions.

_____ 6. When developing a matching test tool, you should do all of the following EXCEPT:
   A. include only one correct match for each item.
   B. arrange statements and responses in random order.
   C. keep all items on the same page of the test booklet.
   D. have the same number of questions and answers.

_____ 7. Which of the following testing tools is NOT considered to be objective?
   A. True/false
   B. Multiple choice
   C. Matching
   D. Essay

_____ 8. Which testing method is the single most important method for determining the competency for actual task performance?
   A. Performance
   B. Written
   C. Oral
   D. Affective

Matching

Match each of the terms in the left column to the appropriate definition.

A. Instructor I responsibility
B. Validity
C. Technical content validity
D. Essay question matrix
E. True/False question
F. Distracters
G. Instructor II responsibility

_____ 1. Test item developed by subject matter expert
_____ 2. Develop and analyze evaluation instruments to ensure that learning objectives are tested
_____ 3. Proctor a written exam to a class
_____ 4. Test item measures what it is intended to measure
_____ 5. Not the correct answers on a multiple-choice exam
_____ 6. Testing tool with a 50/50 chance of getting the answer correct
_____ 7. Assigning relative point values to key responses

True/False

If you believe the statement to be more true than false, write the letter T in the space provided. If you believe the statement to be more false than true, write the letter F.

_____ 1. Essay questions are the easiest to evaluate and grade.
_____ 2. Test questions developed by the instructor are not valid.
_____ 3. Test banks developed by a company are more valid than test banks developed by an instructor.
_____ 4. Testing tools should be developed following APA guidelines.
_____ 5. A written feedback form could be considered a legal document of training performance.
_____ 6. Qualitative analysis is used to determine the acceptability of a question.
_____ 7. Currency of information being asked in a question is as important as is the face value of a question.

Fill-in

Read each item carefully, and then complete the statement by filling in the missing word(s).

1. The purpose of _____ _____ is to determine whether test items are functioning as desired and to eliminate, correct, or modify those test items.

2. The _____ test item is efficient for measuring the application of procedures for starting an apparatus.

Short Answer

Complete this section with short written answers using the space provided.

1. Describe the purpose for analyzing test results.
   _____

Fire Service Instructor I, II

Instructor Applications

Discuss your response to these instructor applications to assist you in developing your knowledge of your responsibilities as a fire service instructor.

1. Review a recent written examination and compare the student responses to the answer key. Identify any test questions that were either too easy (every student answered the question correctly) or that had poor success (there is a high percentage of failures).
   _____

2. Review your local authority policy, and practice posting test scores and student completion records.
   _____

3. Develop a strategy to improve student performance based on a poor evaluation result.
   _____

Evaluating the Fire Service Instructor

Workbook Activities

The following activities have been designed to help you. Your instructor may require you to complete some or all of these activities as a regular part of your instructor training program. You are encouraged to complete any activity that your instructor does not assign as a way to enhance your learning in the classroom.

Fire Service Instructor II

Chapter Review

The following exercises provide an opportunity to refresh your knowledge of this chapter.

Multiple Choice

Read each item carefully, and then select the best response.

_____ 1. What was created during the Wingspread Conference in 1971?
   A. Professional qualifications for fire fighters
   B. Standards for evaluating fire fighters
   C. Fire prevention activities for the fire service
   D. The three E's of fire prevention

_____ 2. When is a fire service instructor evaluated?
   A. During class by another instructor
   B. By the students during a break in the break area
   C. At the end of class during a formal course evaluation
   D. All of the above

_____ 3. (1) Instructor qualifications are set by the authority having jurisdiction. (2) Instructor qualifications are set by the NFPA 1041 standard.
   A. Only statement 1 is correct.
   B. Only statement 2 is correct.
   C. Both statements are correct.
   D. Both statements are incorrect.

_____ 4. Which of the following would NOT be evaluated while doing an evaluation of a new instructor in your training division?
   A. Clothing or attire of the instructor
   B. Occasional references to his or her own department when appropriate
   C. Use of projected media and transitions
   D. Following the lesson plan as outlined

_____ 5. _____ evaluations are intended to refine the instructor's delivery and better prepare him or her to deliver a course.
   A. Summative
   B. Formative
   C. Conclusive
   D. Promotional

CHAPTER 11

_____ 6. Which type of evaluation tool is usually used at the end of the class to determine curriculum strengths and weaknesses?
   A. Summative
   B. Formative
   C. Conclusive
   D. Promotional

_____ 7. Which of the following statements is NOT correct regarding student course evaluations?
   A. Students might use course evaluations to take a shot at the instructor.
   B. Students make comments about course length that are opinions versus constructive criticism.
   C. Course evaluations should be the key source of information for merit increases.
   D. Course evaluation tools should be used to improve the course material.

_____ 8. One of the primary functions of a course evaluation tool is to:
   A. determine students' attitudes about the course.
   B. use the information from the evaluations to discipline the instructor.
   C. justify increasing the budget for the training division.
   D. identify the instructor's weaknesses and then develop a plan to help him or her improve.

_____ 9. Which level of certification requires Fire Service Instructor I as a prerequisite?
   A. Fire Fighter I
   B. Fire Officer I
   C. Fire Investigator
   D. Public Fire and Life Safety-Educator

_____ 10. Which type of evaluation is most often administered by the training division chief?
   A. Formative evaluation
   B. Peer evaluation
   C. Summative evaluation
   D. Departmental evaluation

_____ 11. When evaluating another instructor, the first thing you should do on arriving at the classroom is:
   A. meet with the instructor and let him or her know why you are there and what you will be doing.
   B. slip into the back of the classroom, and make no contact with the instructor you will be evaluating.
   C. visit with students in the class, and ask them how they like the class prior to starting your evaluation.
   D. set up a video camera to record the instructor without his or her knowledge.

_____ 12. During the evaluation of an instructor during training ground activities, the highest priority should be:
   A. ensuring that activities match the lesson plan.
   B. ensuring that safety is always observed.
   C. verifying that the instructor-to-student ratio is within department policy.
   D. ensuring the organization of the training evolutions.

_____ 13. After evaluating an instructor, the next step should be:
   A. leaving as quickly as possible and compiling your notes.
   B. talking with the instructor in front of the class with students still present.
   C. meeting with the instructor one on one and discussing your evaluation.
   D. asking the instructor for his or her thoughts, adding them to your notes, and then leaving.

_____ 14. When discussing your evaluation of an instructor, you should do all of the following EXCEPT:
   A. be objective in your comments.
   B. be honest and friendly.
   C. make some suggestions to improve observed weaknesses.
   D. be blunt and to the point.

_____ 15. All of the following could be measured on an evaluation tool EXCEPT:
   A. whether the class started on time.
   B. whether students liked their classmates.
   C. whether the instructor was properly attired.
   D. whether the room temperature was comfortable.

Matching

Match each of the terms in the left column to the appropriate definition in the right column.

A. NFPA 1500
B. NFPA 1021
C. Fire Service Instructor II
D. Formative evaluation
E. 10
F. NFPA 1041
G. Evaluation tool
H. Strongly agree or disagree
I. Summative evaluation
J. Fire Instructor I

_____ 1. Fire Service Instructor Professional Qualifications standard
_____ 2. Average number of fire fighter deaths during training activities per year (2000–2005)
_____ 3. Improving the fire service instructor's performance
_____ 4. Evaluation completed at the end of a course for information on the course
_____ 5. Evaluation scale found on a student survey
_____ 6. Creates an evaluation tool to evaluate another instructor
_____ 7. Is evaluated by students and supervising instructors
_____ 8. Fire Officer Professional Qualifications standard
_____ 9. Considers attire, eye contact, voice strength, and subject knowledge
_____ 10. Standard on Fire Department Occupational Safety and Health

True/False

If you believe the statement to be more true than false, write the letter T in the space provided. If you believe the statement to be more false than true, write the letter F.

_____ 1. Fire service evaluations are always tied to promotions or merit increases.
_____ 2. On occasion, it is acceptable to deviate from the lesson plan to reinforce a point.
_____ 3. Performing instructor evaluations is for Fire Service Instructor II only.
_____ 4. Fire department policy can exceed the NFPA requirements.
_____ 5. Instructor evaluations are limited to classroom presentation.
_____ 6. Preparation for an evaluation begins with reviewing the lesson plan.
_____ 7. Evaluating an instructor is as important as evaluating the student.
_____ 8. Word of mouth evaluations should be considered an objective source of information.
_____ 9. NFPA 1500 is the basis for all professional qualifications.
_____ 10. Feedback to an instructor should be given with the same confidentiality as that given to a student.

Chapter 11: Evaluating the Fire Service Instructor

Fill-in
Read each item carefully, and then complete the statement by filling in the missing word(s).

1. The lesson plan is the fire service instructor's _____ _____ .

2. Evaluating the _____ _____ is important to student success.

3. During the evaluation, look at the fire service instructor's entire _____ _____ .

4. Closely scrutinizing a fire service instructor's adherence to student _____ guidelines should be paramount.

5. In addition to the _____ of the questions, the _____ of questions asked on an evaluation instrument needs to be considered.

Short Answer
Complete this section with short written answers using the space provided.

1. Describe your department's policy or procedure in respect to fire service instructor evaluations.
   _____

2. Describe the evaluation process.
   _____

3. Describe the role for providing feedback to the fire service instructor.
   _____

Instructor Applications
Discuss your response to these instructor applications to assist you in developing your knowledge of your responsibilities as a fire service instructor.

1. Review the content of a course evaluation form, and identify those elements that relate to fire service instructor qualities as well as those elements that relate to the features of the course.
   _____

2. Develop a list of desirable fire service instructor qualities. Rank these qualities in order based on which create the best learning environment.
   _____

3. Consult available standard personnel evaluations, and identify any features that could be used to evaluate a fire service instructor during a training session.
   _____

Scheduling and Resource Management

Workbook Activities

The following activities have been designed to help you. Your instructor may require you to complete some or all of these activities as a regular part of your instructor training program. You are encouraged to complete any activity that your instructor does not assign as a way to enhance your learning in the classroom.

Fire Service Instructor I, II

Chapter Review

The following exercises provide an opportunity to refresh your knowledge of this chapter.

Multiple Choice

Read each item carefully, and then select the best response.

_____ 1. A critical part of the fire department's efforts to protect its members is the:
  A. training officer/instructor.
  B. fire chief.
  C. company officer.
  D. mayor and city council.

_____ 2. Which of the following is NOT an example of an SOP that must be followed during a training session?
  A. Proper use of PPE
  B. City hall policies
  C. Use of the local ICS
  D. Supervisory responsibilities for safety

_____ 3. Which of the following is NOT an area of consideration when scheduling training?
  A. EMS recertification hours
  B. Hazardous materials OSHA training
  C. CERT refresher training
  D. Department-mandated HR policies

_____ 4. Which type of training takes place at a career fire department?
  A. On the job
  B. Rookie school
  C. Fire academy training
  D. In-service drills

CHAPTER 12

_____ 5. Which of the following is NOT an in-service training category?
   A. Skill/knowledge competence
   B. Skill/knowledge development
   C. Skill/knowledge maintenance
   D. Skill/knowledge improvement

_____ 6. (1) When developing a training schedule, you must consider all areas of the department, such as inspections and technical rescue. (2) Under OHSA, technical rescue teams are responsible for their own skills maintenance and do not follow department policy.
   A. Only statement 1 is correct.
   B. Only statement 2 is correct.
   C. Both statements are correct.
   D. Both statements are incorrect.

_____ 7. When developing a training schedule, you should consider all of the following EXCEPT:
   A. who must attend the training session.
   B. whether this training will impact mutual aid agreements.
   C. who will instruct the training session.
   D. what resources are needed for the training session.

_____ 8. Which type of training would an Instructor I most likely schedule?
   A. Department-level training
   B. Mutual aid training
   C. Drill or company-level training
   D. New equipment training

_____ 9. In general, how many types of training occur in a fire department?
   A. Four
   B. Three
   C. Two
   D. Five

_____ 10. When posting a training notice, you should include all of the following information EXCEPT:
   A. what the training subject is.
   B. what level of PPE or classroom attire is appropriate.
   C. where the training will take place.
   D. what the budget was for the course.

_____ 11. According to NFPA 1500, all training should be based on:
   A. job performance requirements (JPRs).
   B. OSHA regulations.
   C. ISO regulations.
   D. State mandates.

Matching

Match each of the items in the left column with the appropriate description in the right column.

A. NFPA 1500

B. Skill/knowledge improvement

C. Skill/knowledge maintenance

D. Contingency plan B

E. Skill/knowledge development

_____ 1. Process of creating new approach to fire-fighting operations, to be adopted by a department

_____ 2. Training intended to develop performance baselines for core duties and functions

_____ 3. Training intended to correct poor performance or errors observed on the fire ground

_____ 4. Supplemental training schedule

_____ 5. Health and safety standard upon which all training should be based

True/False

If you believe the statement to be more true than false, write the letter T in the space provided. If you believe the statement to be more false than true, write the letter F.

_____ 1. In most departments, it is best not to publish a training schedule that helps to ensure better attendance.

_____ 2. Training requirements for volunteer fire fighters are not the same as for career fire fighters.

_____ 3. The Fire Service Instructor I is responsible for the development of the training budget.

_____ 4. Successful departments have a training schedule that is consistent and easy to understand.

_____ 5. Budget development skills are required at the Instructor I level.

Fill-in

Read each item carefully, and then complete the statement by filling in the missing word(s).

1. Training and education are tools used by fire departments to improve their efficiency in operations and in fire fighter _____.

2. Surveys of training programs often reveal that the delivery of training suffers because of poor _____ of the training _____.

3. Advanced policies may also detail the responsibilities of those who _____, supervised, and _____ in the training.

Short Answer

Complete this section with short written answers using the space provided.

1. Describe how to schedule an instructional session.

_____
_____
_____

Fire Service Instructor II

Multiple Choice
Read each item carefully, and then select the best response.

_____ 1. Training requirements could come from all of the following agencies EXCEPT:
   A. OSHA.
   B. Insurance Service Office (ISO).
   C. Underwriters Lab.
   D. State EMS bureau.

_____ 2. The first step in creating a master training calendar is to:
   A. identify ongoing activities that will affect the training schedule.
   B. identify specialized training needs.
   C. complete an agency needs assessment of regulatory agencies that require training.
   D. compare training requirements proposed by your agency's human resources department.

_____ 3. All of the following are factors seen in a poor training program EXCEPT:
   A. inconsistency of information between fire service instructors.
   B. following of lesson plans and department policies.
   C. failure to adhere or enforce safety practices.
   D. unclear instructions or learning outcomes.

_____ 4. One criterion that outweighs all others in the evaluation of the fire service instructor is:
   A. presentation skills.
   B. adherence to department policy.
   C. safety.
   D. proper instructor-to-student ratio.

Matching
Match each of the items in the left column with the appropriate description in the right column.

A. Agency training needs assessment _____ 1. Basis on which department training is established
B. Annual review and revision _____ 2. External agencies that affect training schedule
C. OSHA, EMS, local SOPs _____ 3. First step in developing a training schedule
D. 5:1 _____ 4. Final step in developing a training schedule
E. Agency training needs assessment _____ 5. Student-to-instructor ratio for high-risk training evolutions

True/False

If you believe the statement to be more true than false, write the letter T in the space provided. If you believe the statement to be more false than true, write the letter F.

_____ 1. Quality assurance is an important aspect of managing a training program.
_____ 2. It is acceptable practice to allow instructors to present lesson plans using their own style.
_____ 3. Mandatory training is required only in OSHA states.
_____ 4. NFPA enforces OSHA.
_____ 5. The master training schedule should be the responsibility of one person.

Fill-in

Read each item carefully, and then complete the statement by filling in the missing word(s).

1. OSHA has specific training requirements that should be considered, covering topics ranging from _____ _____ to _____ _____ _____.

2. A _____ process must be utilized to schedule training sessions.

Short Answer

Complete this section with short written answers using the space provided.

1. Name the external agencies that have an impact on an agency's training schedule.
   _____

2. Describe how to schedule instructors for a training session.
   _____

Fire Service Instructor I, II

Instructor Applications

Discuss your response to these instructor applications to assist you in developing your knowledge of your responsibilities as a fire service instructor.

1. Review your department's training policy, and identify management-related responsibilities that must be completed by the Fire Service Instructor II.
   _____

2. Obtain a copy of the training budget to review expenditures and possible revenue sources.
   _____

3. Review your department's training schedules. Outline the process for determining training subjects, fire service instructor assignments, and the resources needed as part of the overall training program.
   _____

Instructional Curriculum Development

Workbook Activities

The following activities have been designed to help you. Your instructor may require you to complete some or all of these activities as a regular part of your instructor training program. You are encouraged to complete any activity that your instructor does not assign as a way to enhance your learning in the classroom.

Fire Service Instructor III

Chapter Review

The following exercises provide an opportunity to refresh your knowledge of this chapter.

Multiple Choice

Read each item carefully, and then select the best response.

_____ 1. When planning and developing an instructional program, the training officer first determines which of the following priorities?
   A. What the chief wants covered first
   B. What training is necessary
   C. What training the budget allows for
   D. What material was not covered in recruit school

_____ 2. What is at the "center" of the course development and management process?
   A. Cost analysis return on investment
   B. Certification requirements
   C. Training needs assessment
   D. Risk management assessment

_____ 3. How many methods are there to conduct a training needs assessment?
   A. 3
   B. 4
   C. 5
   D. 6

_____ 4. The second step in the four-step teaching model is:
   A. preparation.
   B. evaluation.
   C. application.
   D. presentation.

_____ 5. Which step in the curriculum development process addresses the question "who needs the training (audience)?"
   A. Evaluation step
   B. Preparation step
   C. Application step
   D. Presentation step

CHAPTER 13

_____ 6. The creation and identification of student resources are completed in which step of the CDM?
   A. Presentation step
   B. Preparation step
   C. Evaluation step
   D. Application step

_____ 7. Which step may be considered the most important step of the curriculum development model?
   A. Preparation step
   B. Presentation step
   C. Application step
   D. Evaluation step

_____ 8. A good practice when developing a course goal is to use which of the following methods/models?
   A. CRM
   B. JPR
   C. CDM
   D. ABCD

_____ 9. How often are the NFPA professional qualification standards updated?
   A. Every 3 years
   B. Every 4 years
   C. Every 5 years
   D. Only when they need to be

_____ 10. What is the best method to help remove bias during the review process of vendor-developed curriculum?
   A. Use other instructors to evaluate the material and come to a consensus.
   B. Have the vendors each give a presentation of the material to the chief officers.
   C. Teach both curricula to a group of students and then conduct a vote to see which curriculum is more popular.
   D. Base selection of the curriculum on cost.

_____ 11. Which of the following is the lowest level of learning in the cognitive domain based on Bloom's Taxonomy?
   A. Application
   B. Knowledge
   C. Comprehension
   D. Determination

_____ 12. When a student can select the correct tool for a job and complete the assignment, he or she is operating at which level of Bloom's Taxonomy?
   A. Determination
   B. Knowledge
   C. Comprehension
   D. Application

____ 13. Which of the following agencies specifies or recommends training that an agency must complete in a set year or time frame?
   A. ISO
   B. OSHA
   C. State certification board
   D. All of the above

____ 14. The final step in the curriculum development process is to:
   A. develop the class activities list.
   B. develop the course evaluation tool.
   C. determine what course goals will be taught first.
   D. evaluate instructor qualifications to teach the course being developed.

____ 15. The "condition" of the ABCD model represents which of the following parts of the learning objective?
   A. Who?
   B. How much?
   C. How?
   D. What?

Matching

Match each of the items in the left column with the appropriate description in the right column.

A. JPRs

B. ADDIE

C. Condition

D. Preparation step

E. Job observation

F. Comprehension

G. Method used for assessing needed training based on current personnel

H. Evaluation step

I. Pilot course

J. Degree

____ 1. The most common method to determine training needs for an agency

____ 2. Method used to present new course material to a group of students.

____ 3. Step 1 of the curriculum development process

____ 4. Step 4 of the curriculum development process

____ 5. Another curriculum development model

____ 6. JPR's

____ 7. Level of understanding when a student can clarify or summarize important points

____ 8. Describes the situation in which the student will perform the behavior

____ 9. Firefighter will don an SCBA in 1minute with 100% accuracy

____ 10. Skill and Knowledge Development/Improvement

True/False

If you believe the statement to be more true than false, write the letter T in the space provided. If you believe the statement to be more false than true, write the letter F.

____ 1. Curriculum development is an Instructor II skill.

____ 2. The first step in the curriculum development process is to identify which training is necessary and who needs the training.

____ 3. Course development is driven by national regulations and standards.

____ 4. The curriculum development model used by the NFA involves six steps.

____ 5. The evaluation step of the curriculum development process is when you evaluate which curriculum to use for your agency.

____ 6. A good starting point for writing course goals is to use JPRs.

_____ 7. The use of a vendor-developed curriculum is a good starting point for developing a curriculum for use by your agency.

_____ 8. Knowledge is the highest level of achievement according to Bloom's Taxonomy.

_____ 9. When conducting a pilot course, the best audience is new recruits with no experience in the subject material.

_____ 10. The course goal and end result are the same thing.

Fill-in

Read each item carefully, and then complete the statement by filling in the missing word(s).

1. The NFA has a free, subscription-based resource network titled _____.

2. By following this format (JPRs), the Instructor III can develop _____ _____ for a lesson plan to meet the professional qualifications for the NFPA standards.

3. The _____ aspect of the learning objective describes characteristics of the students.

4. A(n) _____ _____ may be a smaller picture of the desired outcome of the training.

5. By seeking input from _____ _____ regarding training needs, the instructor can focus on training that is relevant and needed by the department and its mission.

Short Answer

Complete this section with short written answers using the space provided.

1. List three methods for conducting a training needs assessment for your agency.
   _____

2. List the four steps used in the curriculum development process.
   _____

3. List the components of a learning objective.
   _____
   _____

Instructor Applications

Discuss your response to these instructor applications to assist you in developing knowledge about your responsibilities as a fire service instructor.

1. Discuss what method is used by your agency to determine training goals and priorities.
   _____

2. Review your agency's policy for curriculum development, and discuss how it could be developed or improved.
   _____

3. Obtain a copy of the TRADE newsletter and review its contents to see if it could become a source for use by your agency.
   _____

Managing the Evaluation System

Workbook Activities

The following activities have been designed to help you. Your instructor may require you to complete some or all of these activities as a regular part of your instructor training program. You are encouraged to complete any activity that your instructor does not assign as a way to enhance your learning in the classroom.

Fire Service Instructor III

Chapter Review

The following exercises provide an opportunity to refresh your knowledge of this chapter.

Multiple Choice Questions

Read each item carefully, and then select the best response.

_____ 1. When should program evaluation tools be used in a course?
   A. At the end of the course
   B. Throughout the entire course
   C. At the beginning and end of the course
   D. Never

_____ 2. A program evaluation tool assesses the level of proficiency for which of the following?
   A. Administrator of the program
   B. Students in the program
   C. Instructors of the program
   D. All of the above

_____ 3. Specific goals and benchmarks set for a specific program must be:
   A. measureable and attainable.
   B. driven by the administration of the program.
   C. focused on the program curriculum.
   D. based on national standards of measurement.

_____ 4. How should program goals be written?
   A. At the end of the course by the instructors
   B. During the course by program administrators, to address the quality of the students and instructors
   C. Before the course starts by both administration and instructors
   D. By the course developer during the curriculum development process

_____ 5. All of the following are commonly used evaluation tools EXCEPT:
   A. student surveys.
   B. strategic surveys.
   C. forecasting surveys.
   D. statistical evaluations.

CHAPTER 14

_____ 6. Which of the following could be considered the "first" component of a program evaluation plan?
   A. Student test scores
   B. Instructor peer evaluations
   C. Student course evaluation tools
   D. Pre- and post-test scores

_____ 7. Which type of grading scale is most often associated with institutes of higher education?
   A. Pass/fail
   B. 75%
   C. A, B, C, or incomplete
   D. 3.5 grade point average

_____ 8. Which of the following concepts will increase the validity of a survey?
   A. Each question converts to a specific goal.
   B. Questions on the survey are written in simple language.
   C. Complete sentences are used for each question.
   D. All of the above

_____ 9. Total scores divided by the number of students will determine the:
   A. mean.
   B. median.
   C. P+ value.
   D. validity.

_____ 10. An instructor should use which of the following calculations to find the true average score?
   A. Mean + Mode
   B. Mean + Median
   C. Mean + Mode + Median
   D. Mode + Median

_____ 11. Which of the following is used to compute the difficulty of a test item?
   A. Mean scores
   B. P+ value
   C. Discrimination index
   D. Reliability index

_____ 12. When developing and evaluating the reliability of test items, the reliability index should be as close to what score as possible?
   A. 1.0
   B. 2.0
   C. .80
   D. .70

_____ 13. Which of the following evaluation tools would be most effective in determining if an instructor presented the lesson objectives as written in a course outline?
   A. Student evaluation
   B. Performance evaluation
   C. Certification test score
   D. Summative evaluation

_____ 14. The last step in a program evaluation process is which of the following?
   A. Adjust test scores as needed due to testing errors.
   B. Set program goals and objectives.
   C. Interpret assessment tool results and reevaluate the program.
   D. Review the program checklist for completion of the program.

_____ 15. How long are test scores or skill sheet results maintained by an agency?
   A. Results are maintained for 30 years.
   B. Length of time depends on whether the agency is within an OSHA state.
   C. Results are maintained based on the agency's policies.
   D. Results are maintained for the service time of the individual plus 30 years.

Matching

Match the following terms with the correct description.

A. Confidentiality
B. Program goal
C. Course outcome
D. Disciplinary procedure
E. Rating scale
F. Mode
G. Median
H. Mean
I. Discrimination index
J. Reliability index

_____ 1. The most commonly occurring score in the examination
_____ 2. Test item's ability to differentiate between test takers' scores
_____ 3. Determined by department policies and procedures
_____ 4. Score located exactly in the middle of the distribution
_____ 5. Designed to be measurable and attainable
_____ 6. Another term for course objective
_____ 7. Total scores divided by number of students
_____ 8. A policy regulating how student test scores and skills evaluations are maintained
_____ 9. A rating of the consistency of results for an overall test
_____ 10. A system of assigning numerical values to survey items

True/False

If you believe the statement to be more true than false, write the letter T in the space provided. If you believe the statement to be more false than true, write the letter F.

_____ 1. Test item analysis is based on pretest analysis.
_____ 2. The Kuder-Richardson formula 21 is used for computing the discrimination index.
_____ 3. The Spearman-Brown formula is used to compute the reliability index of a test item.
_____ 4. A negative "skew" indicates that the majority of students' grades are above the mean score.
_____ 5. Clustering of student scores is a desired result of item analysis.
_____ 6. The first step in post-test item analysis is computing the P-value.
_____ 7. Computing the reliability index is completed after computing the discrimination index.
_____ 8. Each exam item should be reviewed for the level of difficulty.
_____ 9. Results of assessments used during the instructional process should be held until the end of the course and then applied.
_____ 10. The plan of action or strategy should be developed at the same time as the program's goals.

Fill-In

Read each item carefully, and then complete the statement by filling in the missing word(s).

1. When designing a(n) _____ _____, many facets of evaluation and testing will require careful attention to both student results and instructor performance.

2. Evaluation results should be kept _____, and _____ should dictate which staff members and administrative staff have access to the results.

3. Course evaluations are a helpful tool for _____ whether a program has _____ its established goals.

4. A(n) _____ _____ is a listing of each student's answer to a particular question.

5. To identify test items as acceptable or in need of review, the first step is to perform a(n) _____ analysis of the test item.

Short Answer

Complete this section with short written answers using the space provided.

1. Describe the different types of agency policies that relate to program administration and review.
   _____

2. Outline a program evaluation plan.
   _____

3. List the steps in a post-test item analysis process.
   _____

Instructor Applications

Discuss your response to these instructor applications to assist you in developing your knowledge of your responsibilities as a fire service instructor.

1. Review your agency's policies and procedures for record storage, access, and retention.
   _____

2. Evaluate a training program that you currently have in place, and review that evaluation process to determine if it meets the desired goals of the curriculum.
   _____

3. Review your agency's quality improvement program and make recommendations to improve the program as necessary.
   _____

Training Program Management

Workbook Activities

The following activities have been designed to help you. Your instructor may require you to complete some or all of these activities as a regular part of your instructor training program. You are encouraged to complete any activity that your instructor does not assign as a way to enhance your learning in the classroom.

Fire Service Instructor II, III

Chapter Review

The following exercises provide an opportunity to refresh your knowledge of this chapter.

Multiple Choice

Read each item carefully, and then select the best response.

_____ 1. What is the first step in the budget process cycle?
   A. Preparation of a budget request
   B. Review of revenues for the agency
   C. Identification of program needs
   D. Review of budget outlay

_____ 2. Which of the following would be helpful when defending a budget for your training division?
   A. Cost–benefit analysis for your division
   B. Justification summary for each item requested in your budget
   C. Documentation from federal law for items in your budget
   D. None of the following will help in your budget because it's all political anyway.

_____ 3. Fire department budgets are usually based on a(n) _____ cycle.
   A. 6-month
   B. 12-month
   C. 18-month
   D. 24-month

_____ 4. Which of the following would be an example of a capital item for use within a training division?
   A. LCD projector
   B. Printed handouts
   C. Light bill for the facility
   D. Fuel used in the training apparatus

CHAPTER 15

_____ 5. Which of the following would be considered a "consumable" training item?
   A. Textbook
   B. Training tower
   C. Self-contained breathing apparatus
   D. Workbook

_____ 6. When should resources needed to present a lesson be determined?
   A. At the time the lesson is to be presented
   B. As the last step in the lesson plan development process
   C. During the construction of the lesson plan
   D. When gathering materials to present the lesson

_____ 7. Which of the following is the first question to address when determining what type of resources will be purchased for a program?
   A. How will this resource be used in the department?
   B. Which standard will the resource need to meet?
   C. Is this resource compatible with my department's SOP?
   D. Is this resource within the budget limits of my agency?

_____ 8. When determining whether to use a "hand-me-down" chain saw for a training division, which of the following factors is most important?
   A. Currency of the tool
   B. Safety of the saw
   C. Appearance and condition of the saw
   D. Brand and model of the saw

Matching

Match the element of the communications process with its simple definition.

_____ 1. Budget   A. Last step in the budget process
_____ 2. Revenue   B. An example is a burn tower
_____ 3. Expenditures   C. A calendar spanning from the beginning of July to the end of June
_____ 4. Fiscal calendar   D. Income for an agency
_____ 5. Capital expenditure   E. An itemized summary of estimated revenue and expenditures
_____ 6. Training-related expense   F. An example is workbooks and handouts
_____ 7. Monitoring   G. Money spent on salaries, etc.

True/False

If you believe the statement to be more true than false, write the letter T in the space provided. If you believe the statement to be more false than true, write the letter F.

_____ 1. Use of a single source vendor is a violation of the bid process for purchasing equipment.

_____ 2. Training records are private and cannot be used against an agency during a lawsuit.

_____ 3. Developing a training budget is one method in demonstrating the value and impact training has on a department.

_____ 4. Office supplies and fuel for a burn building are considered capital expense.

_____ 5. When using reserve apparatus for training the apparatus does not have to meet NFPA standards since they are being used for training only.

Fill-in

Read each item carefully, and then complete the statement by filling in the missing word(s).

1. Budget preparation is both a(n) _____ and a(n) _____ process.

2. Failure to deliver competent service to a community during an incident may trace its roots back to how the personnel were _____.

3. Budgets must be prepared, justified, and managed throughout the life of the _____ _____.

4. A budget is an itemized summary of estimated _____ and _____.

5. When using reserve apparatus for training make sure _____ is not compromised.

Short Answer

Complete this section with short written answers using the space provided.

1. Identify the usual steps in a budget process.
   _____

2. Identify the questions that need to be addressed when evaluating resources to purchase for a program or curriculum.
   _____

Fire Service Instructor III

Multiple Choice

Read each item carefully, and then select the best response.

_____ 1. Which of the following is considered the most important management function assigned to you as the department training officer?
   A. Developing lesson plans and curriculum
   B. Developing a budget
   C. Selecting instructors to present a lesson
   D. Documenting training through a record-keeping system

_____ 2. When developing a records management system, which NFPA standard should the Instructor III be familiar with?
   A. NFPA 1403
   B. NFPA 1001
   C. NFPA 1401
   D. NFPA 1041

_____ 3. Training records should be maintained in which of the following formats?
   A. Hard copy/paper form
   B. Electronic form
   C. LMS format used by your agency
   D. Record retention is determined by AHJ.

_____ 4. Which of the following is NOT required to be part of a training record file?
   A. The instructor's name
   B. The cost of the course materials
   C. Location where the training took place
   D. Date when the training took place

_____ 5. Who is required to complete records and reports developed during a live fire evolution at an acquired structure?
   A. Fire chief
   B. Incident commander
   C. Instructor in charge
   D. Department training officer

_____ 6. When developing a training division program policy, the first step is to:
   A. determine the purpose and scope of the policy.
   B. determine who will have access to the policy.
   C. determine which person will conduct training on the policy.
   D. determine how to document the training.

_____ 7. In most agencies, how is an instructor selected to present a lesson or topic?
   A. The most senior member of the staff gets to choose what they teach.
   B. The most knowledgeable and experienced instructor is chosen.
   C. The person with the least seniority is chosen.
   D. Selection is based on a rotation.

Matching

Match the element of the communications process with its simple definition.

_____ 1. Educational course   A. An example is a pump operation course taught at the state fire academy

_____ 2. Vocational course   B. An example is a communications course taught at a community college

_____ 3. Who, what, when, where, why   C. Minimum information for a training record

True/False

If you believe the statement to be more true than false, write the letter T in the space provided. If you believe the statement to be more false than true, write the letter F.

_____ 1. A major use of training records is to determine if an agency has met regulatory compliance requirements.
_____ 2. Training records should be complex and difficult to complete.
_____ 3. When completing training records, the lead instructor can initial all areas of the record.
_____ 4. Medical records are not part of a fire fighter's training record.
_____ 5. A progress report of a fire fighter in a training course is not considered part of his or her training record.
_____ 6. Case studies, role-playing, and lectures are all considered delivery methods of training.
_____ 7. Training record retention is mandated by federal law.
_____ 8. NFPA 1401 determines which records and reports are to be created and maintained for live fire training evolutions.
_____ 9. Instructors who are highly skilled in a subject area are also the best instructors to present that topic.

Fill-in

Read each item carefully, and then complete the statement by filling in the missing word(s).

1. If your agency determines that training records are personnel records, then they must be treated as _____ information.

2. The _____ _____ used for training should be identified as part of the training record.

3. Your _____ will determine the delivery method and often the _____ for the program evaluation presentation.

4. Almost every type of training record must identify both the _____ of the course and the _____ who was instructed.

Short Answer

Complete this section with short written answers using the space provided.

1. List the minimum areas for a training record report.
_____

Fire Service Instructor II, III

Instructor Applications

Discuss your response to these instructor applications to assist you in developing your knowledge of your responsibilities as a fire service instructor.

1. Develop a sample budget for a course you are developing. Include instructor pay, textbooks, printed material, and other resources necessary to deliver the course, and then calculate the student tuition that will need to be charged to break even or to stay within department policy.

   _____
   _____
   _____
   _____
   _____

2. Review your department's training schedules, and outline the process for determining training subjects, fire service instructor assignments, and the resources needed as part of the overall training program.

   _____
   _____
   _____
   _____
   _____

3. Review your agency's policy on the selection and assignment of instructors.

   _____
   _____
   _____
   _____
   _____

The Learning Process Never Stops

Workbook Activities

The following activities have been designed to help you. Your instructor may require you to complete some or all of these activities as a regular part of your instructor training program. You are encouraged to complete any activity that your instructor does not assign as a way to enhance your learning in the classroom.

Fire Service Instructor I, II, III

Chapter Review

The following exercises provide an opportunity to refresh your knowledge of this chapter.

Multiple Choice

Read each item carefully, and then select the best response.

_____ 1. Which of the following roles is a prime position to be in to influence the next generation of fire fighters?
  A. Chief
  B. Company officer
  C. Training officer/instructor
  D. None of the above

_____ 2. Which of the following is one method for the fire service instructor to establish his or her creditability?
  A. Teaching a lot of classes
  B. Publishing his or her work in trade journals or books
  C. Attending instructor conferences
  D. Earning a college degree

_____ 3. What sets the fire service leader apart from the average fire fighter?
  A. Having the basic desire to always improve
  B. Always looking for the easy way out of situations
  C. Never volunteering for extra duty or assignments
  D. Following traditions

_____ 4. One of the first steps in becoming a leader in the fire service is:
  A. attending a lot of conferences.
  B. reading as many articles as possible on the Internet or from magazines.
  C. developing an extensive network of friends.
  D. learning time management skills and setting goals.

_____ 5. What is a potential problem or drawback to information found on the Internet?
  A. Cost of Internet access
  B. The accuracy of the information found on the Internet
  C. The time it takes to download information
  D. Copyright issues with material found on the Internet

CHAPTER 16

_____ 6. What is the value of reading articles that contain information from other professions or disciplines outside the fire service?
   A. It might help you find a new profession other than the fire service.
   B. Nothing at all—you should focus on your profession.
   C. It helps to make you a more "well-rounded" person and gives you a more diverse perspective.
   D. It breaks up the routine of reading just fire service material.

_____ 7. What is the educational requirement currently in place to gain entrance into the Executive Fire Officer Program offered by the National Fire Academy?
   A. Bachelor's degree
   B. Associate degree
   C. Master's degree
   D. There is no educational requirement in place.

_____ 8. One of the best reasons to attend major conferences is to:
   A. network and develop friendships.
   B. learn new skills.
   C. take ideas from others and use them as your own.
   D. be away from your work for a period of time.

_____ 9. How many 2-week courses make up the Executive Fire Officer Program?
   A. 3
   B. 5
   C. 4
   D. 2

_____ 10. The process of identifying others to train, mentor, and coach is known as:
   A. succession planning.
   B. replacement training.
   C. progressive retirement.
   D. retooling.

_____ 11. Which of the following is NOT a concept in training your replacement?
   A. Mentoring
   B. Sharing
   C. Coaching
   D. Formalizing

_____ 12. The first step in selecting a person to develop as your replacement is:
   A. mentoring.
   B. identifying.
   C. coaching.
   D. asking.

_____ 13. When mentoring a new protégé, you should do all of the following EXCEPT:
   A. be a sounding board and provide feedback.
   B. be patient and allow time for development.
   C. mold him or her into your image as you see it.
   D. share experiences with your protégé, both good and bad.

_____ 14. The process of letting individuals develop solutions on their own is known as:
   A. mentoring.
   B. coaching.
   C. teaching.
   D. a challenge.

_____ 15. What is something all fire service instructors leave with their departments after they have retired?
   A. Their lesson plans
   B. Their training records
   C. Their attitude
   D. Their legacy of professionalism

Matching
Match the following terms with the correct description or example.

A. Networking
B. ISFSI
C. Mentoring
D. Bachelor's degree
E. FESHE
F. Sharing
G. Publishing an article in a magazine
H. Coaching
I. EFOP
J. National Professional Development Model

_____ 1. Giving to others the "tricks of the trade"
_____ 2. Allowing the protégé to teach a class and then giving him or her feedback
_____ 3. Acting as a sounding board to new ideas and suggestions
_____ 4. National model for degree programs
_____ 5. Four-year program taught at the NFA/NETC in Maryland
_____ 6. National organization that is committed to the development of the instructor
_____ 7. Process of developing contacts with a common interest
_____ 8. EFOP requirement starting in fiscal year 2009 for entrance into program
_____ 9. Process that brings together education, training, and experience
_____ 10. Example of professionalizing the fire service

True/False
If you believe the statement to be more true than false, write the letter T in the space provided. If you believe the statement to be more false than true, write the letter F.

_____ 1. Lifelong learning is a process, not a stopping point.
_____ 2. Information is only as good as its source.
_____ 3. All information found on the Internet is accurate and correct.
_____ 4. Time management is a lifelong skill for professional development.
_____ 5. College education is the only way the fire service can develop into a profession.
_____ 6. Networking is used only for political gain and empowerment.
_____ 7. It is a Fire Service Instructor I and Fire Service Instructor II requirement to develop instructor networks.
_____ 8. The EFOP and the National Professional Development Model are the same thing.
_____ 9. Everyone selected for mentoring works out and succeeds.
_____ 10. Coaching requires positive feedback to be successful.

Fill-in
Read each item carefully, and then complete the statement by filling in the missing word(s).

1. The benefit of reading material outside the fire service profession is that it helps you develop into a more _____-_____ individual and professional.

2. More departments have begun requiring an advanced degree for _____, such as a(n) _____ degree for battalion chief.

3. _____ is a powerful tool in your toolbox that can open the door to sharing or collaborating.

4. Keeping your _____ _____ sharp and current is just as important as reading magazines or attending a conference.

5. The true mark of a great leader is a willingness to plan to be _____.

Short Answer
Complete this section with short written answers using the space provided.

1. Describe the ideas involved in lifelong learning.
   _____

2. Describe the value and importance of time management.
   _____

3. List two or three individuals who could be developed for succession planning for your position.
   _____

Instructor Applications
Discuss your response to these instructor applications to assist you in developing your knowledge of your responsibilities as a fire service instructor.

1. Identify at least three statewide organizations and three national organizations that are available to assist all levels of instructors in their professional development.
   _____

2. Review your local or state certification process to learn which affiliations or correlations exist in accordance with the Fire and Emergency Services Higher Education (FESHE) program discussed in this chapter, "The 16 Firefighter Life-Safety Initiatives" developed by the National Firefighters Foundation, and NFPA standards relating to training and education.
   _____

3. Review the job performance requirements that are specific to your next level of instructor certification, and identify opportunities to experience learning in those areas.
   _____

4. Select an instructor or instructor candidate who has specific talents or background that can be used to improve a particular course. Urge that person to become involved in the training program.
   _____

Fire Service Instructor II

Student Application Package

Developing Objectives, Lesson Plans, and Course Materials

Fire Service Instructor II
Student Application Package

Assignment # _____   Due Date: _____   Pass: _____   Fail: _____

Student Application Package Overview and Guidelines

<u>Purpose:</u> To develop a complete instructional package for presentation and evaluation that meets the job performance requirements of Fire Service Instructor II.

<u>Skill:</u> Manage instructional resources and develop instructional materials used to convey a specific lesson topic to a student audience. (NFPA 1041; 5.2, 5.3, 5.4, 5.5)

<u>Competencies:</u> Prepare and present a complete lesson plan and supporting instructional material to a student audience, completing the following tasks:

- Identify a class subject area for cognitive or psychomotor training.
- Develop three cognitive objectives.
- Develop three psychomotor objectives.
- Prepare a properly formatted lesson plan based on a subject area of choice.
- Develop appropriate media for cognitive objectives presentation.
- Acquire instructional resources for psychomotor objectives presentation.
- Develop an assignment sheet or information sheet on lesson plan information.
- Use a test planning sheet to plan evaluation methods for each objective.
- Develop a psychomotor skills evaluation.
- Develop a written examination using multiple types of questions.
- Develop a class/instructor evaluation form.
- Deliver all prepared materials to a student audience (15 to 20 minutes).

Fire Service Instructor II
Student Application Package

Assignment # _____   Due Date: _____   Pass: _____   Fail: _____

Student Information Sheet for Instructor II

1. Class attendance requirements:

2. Required textbook is *Fire Service Instructor: Principles and Practice, Second Edition* by Jones & Bartlett Learning.

3. Lesson plan material is available from *Fundamentals of Fire Fighter Skills, Third Edition*.

4. Final class projects should be developed using the format suggested in the *Student Applications Guidebook*.

5. The student will develop three cognitive and three psychomotor objectives. Each objective will be in the format of the audience, behavior, condition, degree (ABCD) method as presented in the class.

6. The student will develop a lesson plan on a subject with which the student feels comfortable. Each lesson plan must be based upon a minimum of one objective written for each domain. The lesson plan will be in the format presented in class and will contain all key items as presented in class. Each lesson plan will be limited to 15 to 20 minutes. Students will make two copies of their lesson plan.

7. The student will develop an assignment sheet or an information sheet relating to an objective in her or his cognitive lesson plan.

8. The student will develop an instructor/class presentation evaluation sheet. Copies will be made and distributed to the class while presentations are made. Students will evaluate each other.

9. The student will develop a skill sheet based upon the psychomotor objective used in the lesson plan.

10. Based on the lesson plan objective(s), the student will develop an examination. The examination will contain the following types of questions:

    - Four multiple choice
    - Two true or false
    - Two completion
    - One short answer
    - One matching (must contain a minimum of four items, with one more option in column B than in column A)

Fire Service Instructor II
Student Application Package

Assignment # _____ Due Date: _____ Pass: _____ Fail: _____

Student Information Sheet for Instructor II (Continued)

11. The student will furnish copies of the following for evaluators:

    - Evaluation sheet (student will hand out additional forms to peers)
    - Assignment/information sheet for illustrated lecture
    - Skill sheet for psychomotor lesson
    - Examination
    - Lesson plan

12. The student must use at least one instructional aid during the presentation.

Fire Service Instructor II
Student Application Package

Assignment # _____   Due Date: _____   Pass: _____   Fail: _____

Task: Construct a properly formatted cognitive objective. (NFPA 1041; 5.3.2)

Cognitive objectives involve processing information as part of the learning process. Theories, concepts, and procedures are examples of material that would be included in the cognitive domain.

Complete objectives have four components—Audience, Behavior, Condition, and Degree—abbreviated ABCD. This ABCD method of writing objectives ensures that the critical components of an objective are present. The objective should be written in a statement containing the four components.

AUDIENCE: When writing objectives for a diverse audience, it is important to identify the target members.
  Examples: "The fire officer will . . . ," "The incident commander will . . . ," "Academy candidates will . . . ."

BEHAVIOR: This part of the objective describes the task, activity, knowledge, or attitude being sought.
  Examples: "List classes of fire," "Describe vertical ventilation," "Identify ladder parts."

CONDITION: The objective should state the circumstances under which the outcome will be observed or measured. It should relate as closely as possible to the time limits, materials, or equipment with which the fire fighter will be confronted when performing the task.
  Examples: "given a written exam," "from a selection of pictures," "given a sample lesson plan"

DEGREE: The objective should describe the level or quality of the outcome that, when achieved, indicates acceptable attainment of the task. The standard should reflect acceptable job performance.
  Examples: "within 30 seconds," "with 100% accuracy," "according to department SOP"

Instructions: Using the ABCD method, write three properly formatted cognitive objectives in the spaces provided. Make sure each objective is specific, observable, and measurable.

1. _____

2. _____

3. _____

Fire Service Instructor II
Student Application Package

Assignment # _____   Due Date: _____   Pass: _____   Fail: _____

Task: Construct a properly formatted psychomotor performance objective. (NFPA 1041; 5.3.2)

Psychomotor objectives include the performance of a skill as part of the learning process. *Demonstrate, manipulate,* and *perform* are examples of words that evoke the psychomotor learning domain.

AUDIENCE: When writing objectives for a diverse audience, it is important to identify the target members.
Examples: "The fire officer will . . . ," "The incident commander will . . . ," "Academy candidates will . . . ."

BEHAVIOR: This part of the objective describes the task, activity, or skill.
Examples: "Don SCBA," "Raise the extension ladder," "Demonstrate a straight hose roll."

CONDITION: The objective should state the circumstances under which the outcome will be observed or measured. It should relate as closely as possible to the time limits, materials, or equipment that the fire fighter will be confronted with when performing the task.
Examples: "while wearing full protective clothing," "given a ladder," "given a 1½-, 2½-, or 3-inch hose"

DEGREE: The objective should describe the level or quality of the outcome that, when achieved, identifies acceptable attainment of the task. The standard should reflect acceptable job performance.
Examples: "within 30 seconds," "with 100% accuracy," "according to department SOP"

Instructions: Using the ABCD method, write three properly formatted psychomotor objectives in the spaces provided. Make sure each objective is specific, observable, and measurable.

1. _____
   _____
   _____

2. _____
   _____
   _____

3. _____
   _____
   _____

Fire Service Instructor II

Student Application Package

Assignment # _____ Due Date: _____ Pass: _____ Fail: _____

Task: Develop a properly formatted lesson plan for presentation. (NFPA 1041; 5.3.2)

Instructions: Assemble the components for a lesson plan to present to students in class.

LESSON TITLE: Title hints at the topic and gives the learner some idea of what to expect.
_____

TYPE OF PRESENTATION: Identify whether presentation falls within the cognitive or psychomotor domain.

☐ Cognitive Domain   ☐ Psychomotor Domain

LEARNING OBJECTIVES: Describe what learners will accomplish.

1. _____
   _____

2. _____
   _____

3. _____
   _____

TIME FRAME: Inform the instructor how long it should take to complete the lesson.
_____

LEVEL OF INSTRUCTION: Describe the level of outcome desired from the lesson. Are students expected to know the material to a basic knowledge level or to the analysis level?
_____

MATERIALS NEEDED: List all materials required, including quantity needed, to teach the lesson.
   Examples: number of handouts, video equipment, dry-erase board

_____
_____
_____

Fire Service Instructor II
Student Application Package

Assignment # _____ Due Date: _____ Pass: _____ Fail: _____

Task: Develop a properly formatted lesson plan for presentation. (Continued)

REFERENCES: List the references and resources that were used to develop the lesson. Include page numbers where appropriate.

_____
_____

PREPARATION (step 1): This section reminds the instructor to provide learners with a reason why they need to know this information. How or where will they use this new skill? Motivate the student to pay attention and learn from your presentation.

Examples: Give statistics, refer to case studies and recent incidents, show a video segment.

_____
_____
_____

PRESENTATION (step 2): Outline the lesson. This segment should be in the same order as that in which the information will be presented. Write template in a logical order, step by step, and so forth.

Note: The following is a sample format, your lesson plan may be longer/shorter than space provided.

I.
- A.
- B.
- C.
    1.
    2.

II.
- A.
- B.
    1.
    2.
    3.

III.
- A.
- B.
- C.
- D.

Fire Service Instructor II
Student Application Package

Assignment # _____   Due Date: _____   Pass: _____   Fail: _____

Task: Develop a properly formatted lesson plan for presentation. (Continued)

APPLICATION (step 3): Whenever new information is given or a new skill is taught, it must be followed with an opportunity for the student to apply the new knowledge. Detail what will be used to allow students to apply what was learned.
    Examples: Ask direct questions, provide an exercise, practice the skill, role-play.

_____

_____

_____

_____

EVALUATION (step 4): Evaluate the student's performance. This section should tie into the learning objective. Using the behavior, condition, and standard components of the objective determines how you will measure successful achievement of the objectives. In this segment, you do not need to include your test; only indicate how you will measure the students' performance.
    Examples: written exam, practical skills test, oral exam, written exercise

_____

_____

LESSON SUMMARY: Review the main points of the lesson. This helps to clarify any confusion before dismissal of the class. In most cases, you can review the major topic headings in your class outline.

_____

_____

_____

_____

ASSIGMENTS: State what the student must do to prepare for the next lesson.
    Examples: "Read Chapter X," "Complete the prefire plan," "Practice donning the SCBA five times at your station."

_____

Fire Service Instructor II
Student Application Package

Assignment # _____   Due Date: _____   Pass: _____   Fail: _____

Task: Develop instructional media to support lesson plan and select audiovisual aids. (NFPA 1041; 5.3.2)

Instructions: Develop instructional media to support the delivery of your lesson plan, and identify the audiovisual aids that will be used to deliver your media.

SELECTION OF INSTRUCTIONAL MEDIA: The instructional media should maximize the transfer of knowledge and skills within the time allotted. The instructional media should meet the following criteria:

- Appeals to the senses
- Saves time
- Is relevant to the course objectives
- Is appropriate for the size and interaction of the class
- Is appropriate to the pace of learning
- Has been practiced prior to class

Examples: PowerPoint slides; DVD or CD-ROM; diagrams, charts, graphs; models, props

ASSIGNMENT: From your lesson plan, develop your presentation media to support your delivery of the lesson outline to the class participants.

GENERAL GUIDELINES: If using PowerPoint, the following suggestions may help you complete your presentation:

- Limit the words on your slides to support key points; don't write in sentences.
- Select background, animations, and graphics that do not distract from your presentation.
- Be consistent in font, color, size, and style of your text.
- PRACTICE, PRACTICE, PRACTICE.
- Save your presentation in the correct format for delivery, and make sure all versions of the program are compatible and that you have a backup plan.
  ○ Use a flash drive backed up with a CD-ROM.

Fire Service Instructor II
Student Application Package

Assignment # _____ Due Date: _____ Pass: _____ Fail: _____

Task: Develop a student learning resource for use during practice presentation. (NFPA 1041; 5.3.2)

Instructions: Develop a student learning resource for use during your practice presentation to support the understanding of course objectives.

The following suggestions will guide your development of an information sheet.

CREATE A TITLE OR LIST THE JOB: Indicate the subject area and relate the title to the lesson.

LIST A BEHAVIORAL OBJECTIVE: The objective describes what the information sheet is designed to accomplish.

EXPLAIN THE IMPORTANCE OF THE INFORMATION: Briefly describe the information, and explain how the information will help the student. Motivate the student to read the information.

PRESENT THE INFORMATION: Make the information easy to read. Use charts, graphs, tables, etc., to help explain the information. Use a separate sheet if necessary.

SUMMARIZE THE INFORMATION OR PROVIDE ADDITIONAL RESOURCES/ASSIGNMENTS: Prepare the student for the evaluation step, and make follow-up or additional informational resources available to the students.

> *Examples: copy of a case summary, department SOP, PowerPoint slides handout, fill-in-the-blank or write-in handout, key point summary of terms/concepts*

Fire Service Instructor II
Student Application Package

Assignment # _____ Due Date: _____ Pass: _____ Fail: _____

Task: Develop a skill evaluation sheet for a psychomotor objective. (NFPA 1041; 5.5.2)

Instructions: Develop a student skill evaluation sheet for a psychomotor objective that covers all aspects of the performance skill being taught.

EVALUATION INFORMATION: Your form should include student name, ID number, date, and location, as applicable to your department record-keeping requirements.

LIST THE SKILL (title): Describe in the title what the student will be tested on.
Example: "Soft-Sleeve Hydrant Hook-up"

_____

BEHAVIORAL OBJECTIVE: Every skill evaluation must have the psychomotor objective that is being measured listed in the skill sheet. This provides clarity to the student and instructor on what will be performed, how it will be done, and to what degree the task will be performed.

_____

_____

STUDENT INSTRUCTIONS: Briefly explain the task that the student will perform. Give details of limitations, conditions, and time frames required of the student.

_____

_____

_____

INSTRUCTOR INSTRUCTIONS: Explain to the instructor how the test is to be conducted. Detail any limitations, conditions, time frames, and so forth that the instructor should be aware of.

_____

_____

_____

Fire Service Instructor II

Student Application Package

Assignment # _____   Due Date: _____   Pass: _____   Fail: _____

Task: Develop a skill evaluation sheet for a psychomotor objective. (Continued)

LIST OF THE STEPS TO COMPLETE THE TASK/SKILL: This list consists of the step-by-step procedure of completing the task. Start with the first step in the process and end with the final objective outcome. (You may need to consult manufacturer instructions to complete this step.)

Use the task column in the following table. Expand or reduce the table to fit your skill.

RATING SYSTEM: Determine how the student will be evaluated. If using a scale or point system, provide a description of what is passing.

For a psychomotor skill, the most accepted practice is to use a pass/fail evaluation. The student must complete all steps to be successful. Other measures may help separate performances of students.

- Excellent
- Good
- Average
- Fair
- Poor/Fail

TASK	PASS	FAIL
1.		
2.		
3.		
4.		
5.		

COMMENTS: Provide a place for the instructor to comment on the student's performance. The instructor could make suggestions for improvement if the student failed the test.

_____

_____

_____

SIGNATURE OF EVALUATOR: _____

SIGNATURE OF STUDENT: _____

Fire Service Instructor II
Student Application Package

Assignment # _____   Due Date: _____   Pass: _____   Fail: _____

Task: Develop a written evaluation instrument for course objectives. (NFPA 1041; 5.5.2)

Instructions: Develop a written evaluation for lesson objectives using multiple forms of evaluation measures.

Note: The test planning sheet will be used to help plan the characteristics of your test.

1. WRITE FOUR MULTIPLE CHOICE QUESTIONS: The questions should relate to the class objectives and the lesson plan. *Provide directions for completing this segment.*

   Example: "Choose the correct answer."

   Directions: _____

   1. _____
      A. _____
      B. _____
      C. _____
      D. _____
   2. _____
      A. _____
      B. _____
      C. _____
      D. _____
   3. _____
      A. _____
      B. _____
      C. _____
      D. _____
   4. _____
      A. _____
      B. _____
      C. _____
      D. _____

Fire Service Instructor II
Student Application Package

Assignment # _____ Due Date: _____ Pass: _____ Fail: _____

Task: Develop a written evaluation instrument for course objectives. (Continued)

2. WRITE TWO TRUE/FALSE QUESTIONS: The questions should relate to the class objectives and the lesson plan. *Provide directions for completing this segment.*

   Example: *"Choose the correct answer."*

   Directions: _____

   5. _____
      _____

      A. True

      B. False

   6. _____
      _____

      A. True

      B. False

3. WRITE TWO COMPLETION QUESTIONS: The questions should relate to the class objectives and the lesson plan. *Provide directions for completing this segment.*

   Example: *"Complete the statement."*

   Directions: _____

   7. xxxxxxxxxx xxxxxxxxx xx xxxx xxxxxxxx xx _____ xxxxx xxxx xxxxxxxxxx xxxxxx.

   8. xxxxxxxxxx xxxxxxxxx xx xxxx xxxxxxxx xx _____ xxxxx xxxx xxxxxxxxxx xxxxxx.

4. WRITE ONE SHORT ANSWER QUESTION: The question should relate to the class objectives and the lesson plan. *Provide directions for completing this segment.*

   Example: *"Provide a brief explanation of . . . ."*

   Directions: _____

   9. _____
      _____
      _____
      _____

Fire Service Instructor II
Student Application Package

Assignment # _____ Due Date: _____ Pass: _____ Fail: _____

Task: Develop a written evaluation instrument for course objectives. (Continued)

5. WRITE MATCHING EXERCISE (four items, with one more option in column B than in column A): The question should relate to the class objectives and the lesson plan. *Provide directions for completing this segment.*

Example: "Match the term or item in column A with the appropriate response in column B."

Directions: _____

A	B
10. _____	A. _____
11. _____	B. _____
12. _____	C. _____
13. _____	D. _____
	E. _____

Fire Service Instructor II

Student Application Package

Assignment # _____   Due Date: _____   Pass: _____   Fail: _____

Task: Construct a course and instructor evaluation form. (NFPA 1041; 5.5.3)

Instructions: Develop a course and instructor evaluation form that students can use to review the presentation.

Survey Example: Include instructions for student completion.

    Mark your reaction by circling one of the following choices:
    SA—strongly agree
    A—moderately agree
    D—disagree
    SD—strongly disagree

    SA A D SD   I would take another course like this.
    SA A D SD   I did not learn anything from this class.

Questionnaire Example: Include instructions for student completion.

    Was this course what you expected it to be? If no, why not?

    What areas could be shortened? Entirely eliminated?

Rating Sheet Example: Include instructions for student completion.

    Rating Scale: 5 = outstanding
                             4 = more than satisfactory
                             3 = satisfactory
                             2 = less than satisfactory
                             1 = poor

    Printed materials were well organized.   5 / 4 / 3 / 2 / 1
    Course was a reasonable length.   5 / 4 / 3 / 2 / 1
    Classroom contained minimal distractions. 5 / 4 / 3 / 2 / 1

Fire Service Instructor II
Student Application Package

Assignment # _____   Due Date: _____   Pass: _____   Fail: _____

Instructions: Write your objectives or objective number in the left column, and complete additional column information to ensure that your evaluation instrument is comprehensive.

Test Planning Sheet

Objective Indicate Standard Reference (NFPA or Local)	Objective Type Cognitive or Psychomotor	Evaluation Method <u>Written</u> <u>Practical</u> <u>Other</u> MC Individual Group T/F Team Exercise SA Mult. Obj. Problem Fill-In Essay Matching	Course/ Text/Class Reference	Test Question Number or Practical Evaluation Title

Fire Service Instructor III

Student Application Package

Fire Service Instructor III
Student Application Package

Assignment # _____   Due Date: _____   Pass: _____   Fail: _____

Task: Conduct an agency needs analysis. (NFPA 1041; 6.3.2)
RS: Conducting research, committee meetings, and needs and task analysis. (NFPA 1041; 6.3.2(b))

Conduct an Agency Needs Analysis

Directions: Define the need for the _____ course within your department. Describe the conditions that have led you to the decision that this course is necessary.

1. Identify the motivational needs this course could address when completed.
   _____
   _____

2. Describe the environmental needs this course could address when completed.
   _____
   _____

3. Describe the training needs this course could address when completed.
   _____
   _____

- What methods were used to determine these factors?
  _____
  _____

- What is the impact to service delivery for this course within the community and within the department?
  _____
  _____

Fire Service Instructor III

Student Application Package

Assignment # _____   Due Date: _____   Pass: _____   Fail: _____

Task: Write program and course goals. (NFPA 1041; 6.3.5)
RS: Writing goal statements. (NFPA 1041; 6.3.5(b))

Write Course Goals

Directions: After completing your agency needs assessment and referencing or writing job performance requirements, state your course goal for your course.

Course Name: _____

Course Goal: _____

List terminal objectives. (Terminal objectives are the end results of a part of the training package, also called duty statements.)

1. _____

2. _____

List enabling objectives. (Enabling objectives are the building blocks toward achieving the terminal objective. Multiple enabling objectives may be needed to complete the terminal objective. They become the content outline and performances.)

1a. _____

1b. _____

2a. _____

2b. _____

Fire Service Instructor III
Student Application Package

Assignment # _____   Due Date: _____   Pass: _____   Fail: _____

Task: Write course objectives. (NFPA 1041; 6.3.6)
RS: Writing course objectives and correlating them to job performance requirements (JPRs). (NFPA 1041; 6.3.6(b))

Write JPR

Directions: Using the proper NFPA format, write JPRs for a duty area.

Write Job Heading: _____

Write Duty Heading: _____

1. **Write task statement.** The task statement should include what the person is supposed to do, as taken from the job inventory worksheet. Use an action verb to state the task.
   _____
   _____

2. **List tools, equipment, and materials.** Include items or conditions that must be provided. Answer the "with what?" question.
   _____
   _____

3. **Develop task standard.** How well should this task be done? The standard applies to both performer and evaluator. It promotes consistency in evaluation by reducing the variables used to gauge performance.
   _____
   _____

4. **Identify JPR.** Combine the task statement; description of tools, equipment, materials, and/or conditions of performance; and task standard.
   _____
   _____

5. **Identify prerequisite knowledge and skills.**
   Prerequisite knowledge (basic knowledge that one must have prior to performing the stated task):
   _____

   Prerequisite skill (skills that one must have prior to performing the task):
   _____
   _____

Fire Service Instructor III

Student Application Package

Assignment # _____   Due Date: _____   Pass: _____   Fail: _____

Task: Write course objectives. (NFPA 1041; 6.3.6)
RS: Writing course objectives and correlating them to job performance requirements (JPRs). (NFPA 1041; 6.3.6(b))

Write Objective to Support the JPR

Complete objectives have four components—Audience, Behavior, Condition, and Degree—abbreviated ABCD. This ABCD method of writing objectives ensures that the critical components of an objective are present. The objective should be written in a statement containing the four components.

AUDIENCE: When writing objectives for a diverse audience, it is important to identify the target members.
Examples: "The fire officer will . . . ," "The incident commander will . . . ," "Academy candidates will . . . ."

BEHAVIOR: This part of the objective describes the task, activity, knowledge, or attitude being sought.
Examples: "List classes of fire," "Describe vertical ventilation," "Identify ladder parts."

CONDITION: The objective should state the circumstances under which the outcome will be observed or measured. It should relate as closely as possible to the time limits, materials, or equipment with which the fire fighter will be confronted when performing the task.
Examples: "given a written exam," "from a selection of pictures," "given a sample lesson plan"

DEGREE: The objective should describe the level or quality of the outcome that, when achieved, indicates acceptable attainment of the task. The standard should reflect acceptable job performance.
Examples: "within 30 seconds," "with 100% accuracy," "according to department SOP"

Instructions: Write a complete performance objective that will support your JPR by assembling the four components. Make sure the objective is specific, observable, and measurable.

_____

_____

_____

_____

Fire Service Instructor III
Student Application Package

Assignment # _____   Due Date: _____   Pass: _____   Fail: _____

Task: Select instructional staff. (NFPA 1041; 6.2.4)
RS: Evaluation techniques. (NFPA 1041; 6.2.4(b))

Select Instructional Staff

Directions: Using course information, describe how you will select the instructors for this course.

Instructor prerequisites (as defined by AHJ for certification purposes):

How many instructors will be needed?

- Per day

- Per subject area (Consider practicals requiring additional instructors.)

Department prerequisites:

Other desirable qualifications:

List any instructor selection issues that may arise as a result of this course schedule.

- Compensation:

- Coverage while on duty:

- Consistency in presentation (How will instructors know what to cover and what has been covered?):

- Presentation styles of individual instructors:
  - Technical knowledge (PowerPoint, word processing, etc.):

- Describe instructors' training on how they will teach their portion of the program.

Fire Service Instructor III
Student Application Package

Assignment # _____   Due Date: _____   Pass: _____   Fail: _____

Task: Construct a performance-based instructor evaluation plan. (NFPA 1041; 6.2.5)
RS: Evaluation techniques. (NFPA 1041; 6.2.5(b))

Use this form or develop your own to evaluate instructor performance for your course.

INSTRUCTOR EVALUATION FORM
To be completed by course administrator

Course Title/Number _____

Instructor's name _____

Department _____

Semester/year _____

Use the scale to answer the following questions and make comments. 1 = Strongly disagree, 2 = Disagree, 3 = Somewhat agree, 4 = Agree, 5 = Strongly agree					
Instructor:					
A.	The instructor is prepared for each class.	1	2	3	4	5
B.	The instructor demonstrates knowledge of the subject.	1	2	3	4	5
C.	The instructor provides additional material apart from the textbook.	1	2	3	4	5
D.	The instructor gives citations regarding current situations.	1	2	3	4	5
E.	The instructor communicates the subject matter effectively.	1	2	3	4	5
F.	The instructor shows respect toward students and encourages class participation.	1	2	3	4	5
G.	The instructor maintains an environment conducive to learning.	1	2	3	4	5
H.	The instructor arrives on time.	1	2	3	4	5
I.	The instructor ends class on time.	1	2	3	4	5
J.	The instructor is fair in examination.	1	2	3	4	5

Fire Service Instructor III

Student Application Package

Assignment # _____ Due Date: _____ Pass: _____ Fail: _____

Task: Write equipment purchasing specifications. (NFPA 1041; 6.2.6)
RS: Evaluation methods to select the equipment. (NFPA 1041; 6.2.6(b))

Directions: Using the Resource Evaluation Form, determine the quality, applicability, and compatibility of a training resource for your department.

Table 13-2 Resource Evaluation Form

Criteria	Vendor/Product 1 Rating (1 = poor; 5 = excellent)	Vendor/Product 2 Rating (1 = poor; 5 = excellent)	Vendor/Product 3 Rating (1 = poor; 5 = excellent)

Ease of use within our department
  Instructor
  Student			

Amount of modification needed for our department			

Standard compliance
  NFPA
  OSHA
  ANSI
  Other			

Cost of product/resource			
Similar products already in use by department			
Quantity discount available			
References			
Availability			
Vendor support			
Local alternatives			
Other			

Consider these questions when evaluating your resource:

- How will this resource be used in your department?
- With which standard is the resource developed in conjunction or compliance?
- Is this a turnkey resource or will the instructor need to adapt the resource for local conditions?
- Is this resource compatible with your agency procedures and methods?
- Does the resource fit into your budgetary restrictions?
- Are there advantages to purchasing a larger quantity of the resource?
- Have you received customer testimonials or references for this resource, and if so, do they support your purchase?
- Will the resource be available for use within your prescribed timeline?
- Are there alternatives to purchasing this resource?
- Are manufacturer demonstrations, product experts, and other types of assistance available to the instructor for this resource?

Fire Service Instructor III
Student Application Package

Assignment # _____   Due Date: _____   Pass: _____   Fail: _____

Task: Develop course evaluation plan. (NFPA 1041; 6.5.3)
RS: Decision-making. (NFPA 1041; 6.5.3(b))

Create a program evaluation plan. (NFPA 1041; 6.5.4)
RS: Construction of evaluation instruments. (NFPA 1041; 6.5.4(b))

Create Course Evaluation Plan

Directions: Develop a plan for evaluating your course goal and objectives. Ensure that all objectives are evaluated in some manner.

Objective	Domain	Evaluation		
		Written	Demo	Other

Write a sample test question for one cognitive objective.

Objective:	
Multiple Choice (4 choices)	
True/False	
Matching	
Short Answer	
Essay	

Fire Service Instructor III
Student Application Package

Assignment # _____   Due Date: _____   Pass: _____   Fail: _____

Task: Analyze student evaluation instruments. (NFPA 1041; 6.5.5)
RS: Item analysis techniques. (NFPA 1041; 6.5.5(b))

Complete test item analysis

Directions: Using the test item analysis sheet, determine the reliability of your course test.

Table 14-2 — Test-Item Analysis Tally Sheet

Item Number	Alternatives (Where * indicates the answer)										Responses	
	A	B	C	D	E	F	G	H	I	J	Right	Wrong

Total:

Fire Service Instructor III
Student Application Package

Assignment # _____   Due Date: _____   Pass: _____   Fail: _____

Task: Develop recommendations for policies. (NFPA 1041; 6.2.3)
RS: Technical writing. (NFPA 1041; 6.2.3(b))

Directions: Evaluate the following training issues and determine if they are addressed in your training policies. Identify areas that need improvement. Use the template to recommend policies that may be lacking in your department and forward this information to the fire chief for approval.

Policy	Present? Y/N	Updated (date)	Needs revision? Y/N	Assigned to:
Instructor qualifications				
Training records and documentation				
Temperature extremes				
Medical screening prior to drill				
High-risk drills				
Training equipment				
Non-fire service personnel participation				
Live fire training				
Use of training facility				
Accident or injury during training				

Fire Service Instructor III
Student Application Package

Assignment # _____   Due Date: _____   Pass: _____   Fail: _____

GENERAL ORDER
#2007-2

POLICY NUMBER: _____

SUBJECT: _____ Training _____

EFFECTIVE DATE: _____

AUTHORITY: _____ Chief of Operations _____

REVIEW DATE: _____

REVIEWER: _____ Training Officer _____

Purpose:
_____
_____

Procedure:
_____
_____
_____
_____

Fire Service Instructor III

Student Application Package

Assignment # _____   Due Date: _____   Pass: _____   Fail: _____

Task: **Present evaluation findings, conclusions, and recommendations to agency. (NFPA 1041; 6.2.7)**
RS: Presentation skills and report preparation following agency guidelines. (NFPA 1041; 6.2.7(b))

Administer a training record system. (NFPA 1041; 6.2.2)
RS: Development of forms, report generation. (NFPA 1041; 6.2.2(b))

Develop a system for the acquisition, storage, and dissemination of evaluation results. (NFPA 1041; 6.5.2)
RS: Evaluation, development, and use of information systems. (NFPA 1041; 6.5.2(b))

Present Information on Training Program for Annual Report

Directions: Write a summary of your training program performance for your fire department's annual report. Include the following items. Use graphs where appropriate and quantify performance wherever possible.

- ✓ Significant accomplishments in the area of certifications or major training blocks by personnel
- ✓ Progress toward meeting national standards for training of personnel
- ✓ Purchases of significant training resources made within the budget cycle
- ✓ Number of training hours completed, broken down by rank, type of training, and certifications obtained
- ✓ Future needs to meet training program and agency goals

Answer Key

Chapter 1 Answers

Multiple Choice

1. A   pg. 4
2. D   pg. 4
3. C   pg. 5
4. C   pg. 5
5. D   pg. 6
6. A   pg. 7
7. C   pg. 8
8. D   pg. 10
9. B   pg. 12
10. A   pg. 13
11. C   pg. 13
12. A   pg. 13
13. D   pg. 13
14. A   pg. 19

Matching

1. B   pg. 6
2. B   pg. 7
3. A   pg. 6
4. A   pg. 6
5. C   pg. 7
6. B   pg. 6
7. A   pg. 6
8. B   pg. 7
9. C   pg. 7
10. B   pg. 8
11. C   pg. 9
12. A   pg. 8
13. E   pg. 10
14. D   pg. 10

True/False

1. False   pg. 6
2. True   pg. 7
3. True   pg. 7
4. True   pg. 10
5. True   pg. 11
6. True   pg. 12
7. False   pg. 17
8. False   pg. 15
9. True   pg. 16
10. False   pg. 16

Fill-in

1. professional qualification standards   pg. 5
2. success, failure   pg. 13
3. Standards   pg. 13
4. Codes of ethics   pg. 15
5. accurately recorded   pg. 16

Short Answer

1. Manage the basic resources and the records and reports essential to the instructional process.
   - Assemble course materials.
   - Prepare training records and report forms.

   Review and adapt prepared instructional materials.

   Deliver instructional sessions using prepared course materials.
   - Organize the classroom, laboratory, or outdoor learning environments.
   - Use instructional media and materials to present prepared lesson plans.
   - Adjust presentations to students' different learning styles, abilities, and behaviors.
   - Operate and use audiovisual equipment and demonstration devices.

   Administer and grade student evaluation instruments.
   - Deliver oral, written, or performance tests.
   - Grade students' oral, written, or performance tests.
   - Report test results.
   - Provide examination feedback to students. (pg. 6)

2. Manage instructional resources, staff, facilities, and records and reports.
   - Schedule instructional sessions.
   - Formulate budget needs.
   - Acquire training resources.
   - Coordinate training recordkeeping.
   - Evaluate instructors.

   Develop instructional materials for specific topics.
   - Create lesson plans.
   - Modify existing lesson plans.

   Conduct classes using a lesson plan.
   - Use multiple teaching methods and techniques to present a lesson plan that the instructor has prepared.
   - Supervise other instructors and students during training.

   Develop student evaluation instruments to support instruction and evaluation of test results.
   - Develop student evaluation instruments.
   - Develop a class evaluation instrument.
   - Analyze student evaluation instruments. (pg. 6)

3. Administer agency policies and procedures for training.
   - Administer a training record system.
   - Develop recommendations for policies that support a training program.
   - Select staff of instructors.
   - Construct a performance-based instructor evaluation plan.
   - Write equipment purchasing specifications based on curriculum needs.

   Conduct an agency needs analysis.
   - Design curricula given needs analysis findings and agency goals.
   - Write program and course goals given analysis information and job performance requirements.
   - Modify an existing curriculum to meet the needs of the agency.

   Construct a course content outline given course objectives.
   - Write course objectives that are clear, concise, and measurable.

   Present agency evaluation findings to agency administration.

   Plan, develop, and implement curricula.

   Develop a system for the acquisition, storage, and dissemination of evaluation results.

   Develop a course evaluation plan.

   Analyze student evaluation instruments. (pg. 7)

4. Physical (i.e., lighting, temperature, and setup)

   Emotional (i.e., attitudes, comments, and learning abilities) (pg. 11)
5. Leader, mentor, coach, evaluator, teacher (pg. 8)
6. Leading by example, accountability, recordkeeping, knowledge power base, trust, and confidentiality (pg. 16–17)

Instructor Applications

Answers to these questions will be based on personal experiences and local resources.

Chapter 2 Answers

Multiple Choice

1.	D	pg. 28	6.	D	pg. 33
2.	C	pg. 31	7.	A	pg. 35
3.	A	pg. 31	8.	A	pg. 38
4.	B	pg. 32	9.	C	pg. 33
5.	A	pg. 35	10.	D	pg. 35

Matching

1.	B	pg. 29	9.	B	pg. 31
2.	A	pg. 29	10.	D	pg. 33
3.	C	pg. 29	11.	A	pg. 31
4.	C	pg. 29	12.	C	pg. 38
5.	A	pg. 29	13.	B	pg. 38
6.	B	pg. 29	14.	A	pg. 38
7.	C	pg. 31			
8.	E	pg. 32			

True/False

1.	True	pg. 33	6.	True	pg. 38
2.	False	pg. 33	7.	False	pg. 35
3.	True	pg. 37	8.	True	pg. 35
4.	False	pg. 29	9.	True	pg. 37
5.	True	pg. 39	10.	True	pg. 38

Fill-in

1. ethnic slurs — pg. 32
2. reasonable accommodation — pg. 31
3. sexual advances — pg. 32
4. misfeasance — pg. 35
5. confidentiality — pg. 38
6. social media — pg. 37

Answer Key

Short Answer

1. If injury occurs to a participant or observer during a training session, it is crucial to document what occurred, who was present, and what each person observed. (pg. 38)

2. Personnel files usually include information such as date of birth, social security number, dependent information, and medical information.

    Hiring files include test scores, pre-employment physical reports, psychological reports, and personal opinions about the candidate.

    Disciplinary files include any report or document about an individual's disciplinary history and related reports. (pg. 38)

3. This should include federal law, which is made up of statutory laws; state law, which is established by each state's legislature; and policies and procedures crafted by each individual fire department. (pg. 29)

4. This includes physical or mental impairment that substantially limits one or more major life activities. Major life activities include functions such as caring for oneself, performing manual tasks, walking, seeing, hearing, speaking, breathing, learning, and working. (pg. 31)

5. Works include written words, photographs, and some artwork. (pg. 37)

Instructor Applications

Answers to these questions will be based on personal experiences and local resources.

Chapter 3 Answers

Multiple Choice

1. A	pg. 52	7. C	pg. 58	13. B	pg. 55
2. C	pg. 52	8. A	pg. 58	14. C	pg. 57
3. A	pg. 52	9. B	pg. 59	15. D	pg. 62
4. B	pg. 53	10. C	pg. 66		
5. D	pg. 53	11. B	pg. 67		
6. B	pg. 53	12. B	pg. 67		

Matching

1. C	pg. 54	8. D	pg. 65
2. B	pg. 54	9. A	pg. 65
3. D	pg. 54	10. F	pg. 65
4. A	pg. 54	11. B	pg. 65
5. B	pg. 52	12. E	pg. 65
6. C	pg. 52	13. C	pg. 65
7. A	pg. 53		

True/False

1. True	pg. 52	7. True	pg. 54
2. False	pg. 66	8. True	pg. 59
3. True	pg. 67	9. False	pg. 66
4. True	pg. 63	10. True	pg. 65
5. True	pg. 58	11. True	pg. 55
6. False	pg. 53	12. False	pg. 55

Fill-in

1. motivation — pg. 53
2. generation X — pg. 58
3. short pause — pg. 64
4. 10 percent, 90 percent — pg. 65
5. formal education — pg. 57

Short Answer

1. Motivational techniques include building on the adult learner's reason for being in class, making attendance mandatory, expressing why the students must know the information, and explaining how the student will learn the material. (pg. 53)

2. The classroom should be distraction free and centered on student participation and comfort, and lessons should be easy to see and hear. (pg. 54)

3. Effective learning is a natural process, there are three types of learners, motivation is a key factor, and there are many similarities and differences in learners. (pg. 55–56)

4. To deal with the monopolizer, make certain to call on other students during classroom discussions.

    With the historian, carefully guide the discussion back to the main topic. Another way to redirect real historians is by pairing them up with struggling students. Artificial historians are almost always distracting. Assertively redirecting the discussion back to the lesson plan generally corrects this issue. If not, take the student aside and discuss the need for the class to stay focused on the lesson plan.

    For the day dreamer, try to draw such students into the class by using direct questioning techniques and making maximum eye contact.

    For the expert, you must corral the expert and may even have to discuss his or her contributions to the class during a private break. (pg. 66–67)

5. Baby boomers involve senior members with experience and knowledge. Generation X members have many questions and seek instant gratification. Generation Y members seem spoiled, with little mechanical but great comprehensive abilities. (pg. 57–61)

Instructor Applications

Answers to these questions will be based on personal experiences and local resources.

Chapter 4 Answers

Multiple Choice

1. D	pg. 79	9. A	pg. 86
2. C	pg. 79	10. B	pg. 86
3. A	pg. 86	11. A	pg. 86
4. B	pg. 79	12. D	pg. 86
5. D	pg. 79	13. C	pg. 87
6. A	pg. 79	14. B	pg. 86
7. C	pg. 79	15. C	pg. 79
8. C	pg. 86		

Answer Key

Matching

1. B   pg. 86
2. C   pg. 86
3. A   pg. 86
4. H   pg. 86
5. F   pg. 86
6. D   pg. 86
7. E   pg. 86
8. G   pg. 86
9. I   pg. 86
10. M pg. 91
11. K pg. 91
12. L pg. 91
13. J pg. 91

True/False

1. True   pg. 78
2. True   pg. 86
3. False pg. 90
4. False pg. 87
5. False pg. 90
6. True   pg. 91
7. True   pg. 87
8. True   pg. 86
9. True   pg. 86
10. False pg. 90

Fill-in

1. Learning   pg. 79
2. time, patience   pg. 79
3. learning style   pg. 89
4. Taxonomy of Learning pg. 86
5. readiness   pg. 79

Short Answer

1. The law of readiness: A person can learn why physically and mentally he or she is ready to respond to instruction.

    The law of exercise: Learning is an active process that exercises both the mind and the body.

    The law of effect: Learning is most effective when it is accompanied by or results in a feeling of satisfaction, pleasantness, or reward (internal or external) for the student.

    The law of association: In the learning process, the learner compares the new knowledge with his or her existing knowledge base.

    The law of recency: Practice makes perfect, and the more recent the practice, the more effective the performance of the new skill or behavior.

    The law of intensity: Real-life experiences are more likely to produce permanent behavioral changes, making this type of learning very effective (pg. 78).

2. Cognitive—knowledge, psychomotor—physical use of knowledge, affective—attitudes, emotions, or values (pg. 86)

3. Each student has a different learning style—that is, a way in which he or she learns most effectively (pg. 89).

Instructor Applications

Answers to these questions will be based on personal experiences and local resources.

Chapter 5 Answers

Multiple Choice

1.	D	pg. 105	6.	A	pg. 106	11.	A	pg. 110
2.	B	pg. 105	7.	C	pg. 110	12.	C	pg. 112
3.	A	pg. 105	8.	D	pg. 107	13.	D	pg. 115
4.	C	pg. 105	9.	B	pg. 115	14.	B	pg. 105
5.	D	pg. 105	10.	A	pg. 107	15.	C	pg. 109

Matching

1. D pg. 105
2. E pg. 105
3. C pg. 105
4. A pg. 105
5. B pg. 105

True/False

1.	True	pg. 105	6.	False	pg. 110
2.	True	pg. 107	7.	True	pg. 115
3.	True	pg. 104	8.	False	pg. 110
4.	False	pg. 107	9.	True	pg. 107
5.	True	pg. 107	10.	True	pg. 109

Fill-in

1. enthusiasm, enthusiasm   pg. 109
2. 10, 90   pg. 106
3. active, passive   pg. 107
4. communicator   pg. 104
5. informative speeches, lectures   pg. 114
6. technical in nature   pg. 112

Short Answer

1. The basic communication process consists of five elements: the sender, the message, the medium, the receiver, and feedback. (pg. 107)

2. Communication is the most important element of the learning process. It takes many formal and informal forms. Instructors guide its flow and pace. (pg. 106)

3. Verbal includes oral communication. (pg. 105) Nonverbal includes the tone of voice, eye movement, posture, hand gestures, and facial expressions. (pg. 105) Written communication is used to document both routine and extraordinary fire department activities. (pg. 110)

Instructor Applications

Answers to these questions will be based on personal experiences and local resources.

Chapter 6 Answers

Fire Service Instructor I

Multiple Choice

1. A	pg. 127		9. B	pg. 132
2. D	pg. 126		10. A	pg. 132
3. D	pg. 126		11. C	pg. 134
4. D	pg. 127		12. C	pg. 134
5. C	pg. 128		13. B	pg. 136
6. D	pg. 128		14. A	pg. 132
7. A	pg. 127		15. B	pg. 132
8. B	pg. 128			

Matching

1. G	pg. 132		7. C	pg. 127
2. H	pg. 132		8. B	pg. 131
3. A	pg. 130		9. C	pg. 134
4. D	pg. 132		10. B	pg. 134
5. E	pg. 132		11. A	pg. 133
6. F	pg. 132		12. D	pg. 134

True/False

1. True	pg. 128		6. False	pg. 129
2. True	pg. 132		7. True	pg. 126
3. False	pg. 136		8. True	pg. 127
4. True	pg. 136		9. True	pg. 134
5. False	pg. 132		10. False	pg. 134

Fill-in

1. alter, lesson objectives   pg. 136
2. Instructor I   pg. 133
3. learning objective   pg. 127
4. level of instruction   pg. 131
5. adapt, modify   pg. 135

Short Answer

1. Audience, behavior, condition, degree (pg. 127)
2. Lesson title, level of instruction, behavioral objectives, instructional materials, lesson outline, references, lesson summary, assignment (pg. 129–132)
3. Preparation, presentation, application, evaluation (pg. 132–134)
4. Instructor I can adapt a lesson plan to meet the needs of an audience or local conditions. Instructor II can make basic changes to the lesson plan, such as rewriting learning objectives or performance outcomes. (pg. 136)

Instructor Applications

Answers to these questions will be based on personal experiences and local resources.

Fire Service Instructor II

Multiple Choice

1. B   pg. 141
2. D   pg. 140
3. C   pg. 140
4. A   pg. 140
5. B   pg. 140
6. D   pg. 141
7. C   pg. 141
8. A   pg. 141
9. B   pg. 142–143
10. C   pg. 143
11. D   pg. 143
12. A   pg. 143
13. C   pg. 144
14. B   pg. 145
15. D   pg. 149

Matching

1. H   pg. 140
2. B   pg. 140
3. D   pg. 140
4. I   pg. 140
5. F   pg. 141
6. A   pg. 143
7. E   pg. 143
8. J   pg. 144
9. C   pg. 145
10. G   pg. 141

True/False

1. True   pg. 140
2. True   pg. 140
3. False   pg. 141
4. False   pg. 140
5. False   pg. 141
6. True   pg. 141
7. True   pg. 143
8. False   pg. 143
9. True   pg. 148
10. True   pg. 148

Fill-In

1. template, consistency   pg. 140
2. clarifying   pg. 140
3. lesson outline   pg. 145
4. evaluation, learning   pg. 148
5. validity   pg. 149

Short Answer

1. Cognitive, psychomotor, affective (pg. 141–143)

Instructor Applications

Answers to these questions will be based on personal experiences and local resources.

Chapter 7 Answers

Multiple Choice

1. A	pg. 169	9. C	pg. 166
2. B	pg. 169	10. A	pg. 163
3. C	pg. 170	11. B	pg. 169
4. D	pg. 164	12. D	pg. 165
5. A	pg. 164	13. A	pg. 163
6. D	pg. 161	14. C	pg. 167
7. C	pg. 165	15. B	pg. 161
8. C	pg. 166		

Matching

1. H	pg. 160	6. J	pg. 163
2. C	pg. 169	7. A	pg. 163
3. B	pg. 164	8. D	pg. 166
4. E	pg. 164	9. I	pg. 165
5. G	pg. 164	10. F	pg. 169

True/False

1. False	pg. 162	6. True	pg. 170
2. True	pg. 162	7. False	pg. 164
3. True	pg. 169	8. False	pg. 165
4. False	pg. 170	9. False	pg. 164
5. False	pg. 170	10. True	pg. 163

Fill-in

1. culture — pg. 161
2. formal, behavior — pg. 163
3. contingency — pg. 169
4. personnel, safety — pg. 167
5. grin, in — pg. 170

Short Answer

1. Demographics include race, national origin, age, gender, marital status, family size, and educational background. Other possible factors are type of fire agency, career, volunteer, or paid-on-call. (pg. 169)

2. Learning environment consists of the classroom, lesson plans, lesson outlines, and an evaluation tool. (pg. 163)

3. A student should be comfortable in regard to classroom temperature, there should be a sufficient number of breaks during instruction, and the learning environment should be safe, should encourage teamwork but allow for mistakes, and should allow the student to be successful. (pg. 169)

Instructor Applications

Answers to these questions will be based on personal experiences and local resources.

Answer Key   127

Chapter 8 Answers

Multiple Choice

1. A	pg. 185	9. A	pg. 188
2. B	pg. 187	10. B	pg. 194
3. C	pg. 181	11. C	pg. 195
4. B	pg. 189	12. D	pg. 189
5. C	pg. 191	13. A	pg. 197
6. A	pg. 189	14. B	pg. 197
7. C	pg. 193	15. C	pg. 197
8. D	pg. 193		

Matching

1. G	pg. 188	6. F	pg. 188
2. E	pg. 188	7. A	pg. 179
3. I	pg. 195	8. H	pg. 185
4. B	pg. 191	9. D	pg. 195
5. C	pg. 193	10. J	pg. 196

True/False

1. True	pg. 197	6. True	pg. 188
2. False	pg. 196	7. True	pg. 189
3. True	pg. 197	8. False	pg. 190
4. False	pg. 190	9. True	pg. 194
5. False	pg. 190	10. True	pg. 196

Fill-in

1. variety — pg. 188
2. over-reliance — pg. 189
3. scanner — pg. 190
4. digital audio player — pg. 194
5. simulator — pg. 196

Short Answer

1. Understand the equipment that you will use during a presentation, and practice using it before your classroom presentation. (pg. 188)
2. Check the connections between the computer and the projector, check the projector's light bulb, and make sure the cover is off the lens of the projector. (pg. 198)
3. Multimedia appeals to more senses, thereby increasing the retention of learning and keeping the learner involved in the learning process. (pg. 197)

Instructor Applications

Answers to these questions will be based on personal experiences and local resources.

Chapter 9 Answers

Multiple Choice

1.	A	pg. 208	9.	A	pg. 218
2.	B	pg. 208	10.	B	pg. 218
3.	C	pg. 209	11.	C	pg. 218
4.	D	pg. 210	12.	B	pg. 219
5.	A	pg. 210	13.	C	pg. 210
6.	B	pg. 210	14.	C	pg. 220
7.	C	pg. 211	15.	D	pg. 220
8.	D	pg. 215			

Matching

1.	F	pg. 208	6.	E	pg. 221
2.	I	pg. 209	7.	J	pg. 215
3.	G	pg. 210	8.	B	pg. 218
4.	C	pg. 210	9.	A	pg. 214
5.	D	pg. 210	10.	H	pg. 215

True/False

1.	True	pg. 208	6.	True	pg. 211
2.	False	pg. 208	7.	False	pg. 215
3.	True	pg. 210	8.	False	pg. 216
4.	True	pg. 210	9.	True	pg. 218
5.	False	pg. 211	10.	True	pg. 218

Fill-in

1. empowered — pg. 209
2. technology, safety — pg. 209
3. creative — pg. 216
4. personnel — pg. 216
5. experience, drill ground — pg. 219

Short Answer

1. Departments should use the 16 Life-Safety Initiatives in their polices and procedures to change the culture of the fire department to make it safer. (pg. 208–209)

2. You teach safety as an instructor by example. Follow department policy in all of your training activities. Do not allow unsafe activities from those you train or those who assist you in the training process. (pg. 210–215)

3. Safety is an attitude that becomes a part of you and how you do business. A good instructor learns to anticipate problems before they happen; this knowledge comes from fire-ground experience, training, and good preparation. (pg. 219)

Instructor Applications

Answers to these questions will be based on personal experiences and local resources.

Chapter 10 Answers

Fire Service Instructor I, II

Multiple Choice

1. A   pg. 231
2. B   pg. 231
3. C   pg. 231
4. B   pg. 234
5. C   pg. 234
6. D   pg. 232
7. C   pg. 234

Matching

1. B   pg. 231
2. A   pg. 231
3. C   pg. 231

True/False

1. False   pg. 232
2. False   pg. 234
3. True   pg. 234

Fill-in

1. testing   pg. 230
2. written test   pg. 231
3. policy   pg. 234

Short Answer

1. Lack of standardized test specifications and test format. Confusing procedures and guidelines for test development, review, and approval. Lack of consistency and standardization in the application of testing technology. Failure to perform formal test-item analysis. Inadequate fire service instructor training in testing methods (pg. 230).

2. Face validity, technical content validity, job content validity or criterion reference validity, and currency of the information (pg. 237).

Fire Service Instructor II

Multiple Choice

1. D   pg. 238
2. A   pg. 237
3. A   pg. 238
4. B   pg. 240
5. C   pg. 240
6. D   pg. 242
7. D   pg. 245
8. A   pg. 247

Matching

1. C   pg. 237
2. G   pg. 238
3. A   pg. 236
4. B   pg. 238
5. F   pg. 240
6. E   pg. 244
7. D   pg. 247

True/False

1. False   pg. 247
2. False   pg. 239
3. True   pg. 239
4. True   pg. 235
5. True   pg. 234
6. True   pg. 237
7. True   pg. 237

Fill-in

1. test analysis   pg. 237
2. arrangement   pg. 242

Fire Service Instructor I, II

Short Answer

1. To evaluate the results to see that course objectives were met and that the questions measured the correct information (pg. 237).

Instructor Applications

Answers to these questions will be based on personal experiences and local resources.

Chapter 11 Answers

Multiple Choice

1. A   pg. 260
2. D   pg. 260
3. C   pg. 260-261
4. B   pg. 261
5. B   pg. 262
6. A   pg. 262
7. C   pg. 264
8. D   pg. 267
9. B   pg. 262
10. C   pg. 262
11. A   pg. 265
12. B   pg. 265
13. C   pg. 267
14. D   pg. 267
15. B   pg. 270

Matching

1. F   pg. 260
2. E   pg. 261
3. D   pg. 262
4. I   pg. 262
5. H   pg. 270
6. C   pg. 269
7. J   pg. 263
8. B   pg. 262
9. G   pg. 262
10. A   pg. 267

True/False

1. False   pg. 262
2. True   pg. 265
3. True   pg. 261
4. True   pg. 261
5. False   pg. 265
6. True   pg. 265
7. True   pg. 262
8. False   pg. 269
9. False   pg. 260
10. True   pg. 267

Fill-in

1. road map   pg. 265
2. learning environment pg. 270
3. skill set   pg. 264
4. safety   pg. 265
5. format, number   pg. 271

Short Answer

1. Policy should contain the evaluation tool or explain how to obtain or develop one. The policy should address when evaluations should be given, by whom, and for what purpose and should then explain what to do once the evaluation is complete, including feedback to the instructor. (pg. 261–262)

2. Before evaluating an instructor, you should review the lesson plan and content, obtain the evaluation tool, and review its content. Review department policy for conducting an evaluation. Determine the purpose of the evaluation and follow department policy regarding the reason for the evaluation. Arrive early, introduce yourself to the instructor, and let him or her know the reason you are in his or her class. Complete the evaluation form during the allotted time. After the form is completed, follow your department's policy for providing feedback to the instructor. (pg. 264)

3. After you have evaluated an instructor, you should provide feedback to him or her. In doing so, you should key in on the positives first and then, as appropriate, identify areas that need to be improved. When offering suggestions, do so in a friendly yet firm manner and offer any ideas on how to improve and be supportive. This is an opportunity to coach an instructor to make him or her better and increase his or her skills in teaching the next generation of fire fighters. (pg.266–269)

Instructor Applications

Answers to these questions will be based on personal experiences and local resources.

Chapter 12 Answers

Fire Service Instructor I, II

Multiple Choice

1. A   pg. 280
2. B   pg. 280
3. C   pg. 280–281
4. D   pg. 282
5. A   pg. 282, 285
6. A   pg. 280
7. B   pg. 280
8. C   pg. 281
9. C   pg. 282
10. D   pg. 287
11. A   pg. 282

Matching

1. E   pg. 282
2. C   pg. 282
3. B   pg. 285
4. D   pg. 285
5. A   pg. 282

True/False

1. False   pg. 286
2. False   pg. 281
3. False   pg. 287
4. True   pg. 280
5. False   pg. 280

Fill-in

1. safety   pg. 280
2. management, schedule   pg. 281
3. attended, participated   pg. 287

Short Answer

1. Determine what is to be taught, who your audience is, and where and when the training should take place. Arrange for a qualified instructor, and have the training approved if necessary. Send out notice to those who will be required to attend, and ensure that all necessary resources will be available. (pg. 286–287)

Fire Service Instructor II

Multiple Choice

1. C   pg. 289
2. C   pg. 290
3. B   pg. 292
4. C   pg. 290

Matching

1. E   pg. 290
2. C   pg. 289
3. A   pg. 287
4. B   pg. 290
5. D   pg. 289

True/False

1. True   pg. 292
2. True   pg. 290
3. False   pg. 289
4. False   pg. 289
5. True   pg. 289

Fill-in

1. respiratory protection, infectious disease protection   pg. 289
2. standardized   pg. 289

Short Answer

1. The following agencies have an impact on your training schedule, either directly or indirectly: OSHA, ISO, NFPA, state fire marshal's office, state or national emergency services system, state driver's license authority, and state fire service certification programs. (pg. 289)

2. Select an instructor who is qualified as an instructor and has a knowledge base of the subject to be taught. Once the instructor is approved, supply him or her with lesson material and resources for the course to be taught. (pg. 289)

Fire Service Instructor I, II

Instructor Applications

Answers to these questions will be based on personal experiences and local resources.

Chapter 13 Answers

Multiple Choice

1. B   pg. 300
2. C   pg. 300
3. A   pg. 301
4. D   pg. 304
5. B   pg. 304
6. A   pg. 305
7. C   pg. 305
8. D   pg. 307
9. C   pg. 307
10. A   pg. 309
11. B   pg. 311
12. D   pg. 311
13. D   pg. 301
14. B   pg. 305
15. C   pg. 310

Answer Key

Matching

1. E   pg. 301
2. I   pg. 313
3. D   pg. 304
4. H   pg. 305
5. B   pg. 303
6. A   pg. 307
7. F   pg. 311
8. C   pg. 311
9. J   pg. 311
10. G   pg. 303

True/False

1. False   pg. 300
2. True   pg. 300
3. True   pg. 301
4. False   pg. 303
5. False   pg. 305
6. True   pg. 306
7. True   pg. 307
8. False   pg. 311
9. False   pg. 314
10. True   pg. 304

Fill-in

1. TRADE   pg. 314
2. learning objectives   pg. 312
3. audience   pg. 310
4. course goal   pg. 306
5. line officers   pg. 301

Short Answer

1. Job observation, compliance and regulation, skill/knowledge development and improvement (pg. 301)

2. Step 1—Preparation (assembling course material), Step 2—Presentation (materials used to deliver the course), Step 3—Application (activities used to enhance learning and engage the learner), Step 4—Evaluation (materials used to evaluate the learning process) (pg. 304–305)

3. ABCD model: A—Audience, B—Behavior, C—Condition, D—Degree (pg. 310)

Instructor Applications

Answers to these questions will be based on personal experiences and local resources.

Chapter 14 Answers

Multiple Choice

1. B   pg. 325
2. D   pg. 327
3. A   pg. 327
4. C   pg. 327
5. B   pg. 328
6. C   pg. 328
7. D   pg. 330
8. D   pg. 332
9. A   pg. 336
10. C   pg. 337
11. B   pg. 337
12. A   pg. 339
13. B   pg. 339
14. C   pg. 340
15. C   pg. 324

Answer Key

Matching

1. F   pg. 336
2. I   pg. 337
3. D   pg. 328
4. G   pg. 336
5. B   pg. 327
6. C   pg. 327
7. H   pg. 336
8. A   pg. 324
9. J   pg. 338
10. E   pg. 330

True/False

1. False   pg. 332
2. False   pg. 338
3. True   pg. 338
4. True   pg. 336
5. True   pg. 335
6. False   pg. 334
7. True   pg. 334
8. True   pg. 334
9. False   pg. 327
10. True   pg. 328

Fill-In

1. evaluation system   pg. 324
2. confidential, policy   pg. 325
3. assessing, achieved   pg. 328
4. item analysis   pg. 332
5. quantitative   pg. 339

Short Answer

1. Agency policies include administrative titles and duties, disciplinary procedures, records and reporting procedures, flowchart or area of responsibility diagrams, teaching curriculum, quality improvement measures, grading policies, and code of conduct for students and instructors. (pg. 328)

2. A program evaluation plan would include course evaluations completed by the students, instructor evaluations, test results, agency policies and procedures, facilities and equipment survey, and safety review and analysis. (pg. 328)

3. Gather data, compute the P+ value, compute the discrimination index, compute the reliability index, identify whether items are acceptable or need review, and revise test items based on post-test analysis. (pg. 334)

Instructor Application

Individual answers will depend on department operations and individual beliefs and impressions.

Chapter 15 Answers
Fire Service Instructor II, III

Multiple Choice

1. C   pg. 350
2. B   pg. 351
3. B   pg. 351
4. A   pg. 353
5. D   pg. 353
6. C   pg. 355
7. A   pg. 356
8. B   pg. 354

Matching

1. E   pg. 350
2. D   pg. 350
3. G   pg. 350
4. C   pg. 351
5. B   pg. 353
6. F   pg. 353
7. A   pg. 353

True/False

1. False   pg. 355
2. False   pg. 350
3. True   pg. 351
4. False   pg. 353
5. False   pg. 354

Fill-in

1. technical, political   pg. 351
2. trained   pg. 350
3. budget cycle   pg. 353
4. revenues, expenditures   pg. 350
5. safety   pg. 355

Short Answer

1. Identification of needs and required resources, preparation of a budget request, local government and public review of requested budget, adoption of an approved budget, administration of approved budget, close-out of budget year (pg. 350–351)

2. How will this product/resource be used in your department? Which standard is the product/resource developed in conjunction or compliance with? Is this a "turnkey" product? Is this product/resource compatible with your agency's procedures and methods? Does the product/resource fit within budgetary restrictions? Are there advantages to purchasing larger quantities of the product/resource? Can you get references from vendors for the product? Can the product/resource be obtained in a timely manner? Can the product/resource be found locally? (pg. 356)

Fire Service Instructor III

Multiple Choice

1. D   pg. 359
2. C   pg. 359
3. D   pg. 359–360
4. B   pg. 365
5. C   pg. 365
6. A   pg. 366
7. B   pg. 369

Matching

1. B   pg. 363
2. A   pg. 363
3. C   pg. 365

True/False

1. False   pg. 361
2. False   pg. 361
3. True   pg. 365
4. False   pg. 355
5. False   pg. 360
6. True   pg. 363
7. True   pg. 360
8. False   pg. 365
9. False   pg. 369

Fill-in

1. confidential   pg. 360
2. delivery method   pg. 363
3. audience, medium   pg. 369
4. instructor, person   pg. 360

Short Answer

1. Who, what, when, where, and why (pg. 365)

Fire Service Instructor II, III

Instructor Applications

Answers to these questions will be based on personal experiences and local resources.

Chapter 16 Answers

Multiple Choice

1. C	pg. 382	9. C	pg. 388
2. B	pg. 382	10. A	pg. 388
3. A	pg. 383	11. D	pg. 388
4. D	pg. 383	12. B	pg. 390
5. B	pg. 384	13. C	pg. 391
6. C	pg. 384	14. B	pg. 391
7. A	pg. 384	15. D	pg. 391
8. A	pg. 385		

Matching

1. F	pg. 391	6. B	pg. 386
2. H	pg. 391	7. A	pg. 385
3. C	pg. 390	8. D	pg. 384
4. E	pg. 388	9. J	pg. 388
5. I	pg. 388	10. G	pg. 382

True/False

1. True	pg. 382	6. False	pg. 385
2. True	pg. 383	7. False	pg. 386
3. False	pg. 384	8. False	pg. 388
4. True	pg. 383	9. False	pg. 390
5. False	pg. 384	10. True	pg. 391

Fill-in

1. well-rounded — pg. 384
2. promotion, bachelor's — pg. 384
3. Networking — pg. 385
4. teaching skills — pg. 386
5. replaced — pg. 388

Short Answer

1. Lifelong learning is the basic idea that you should always be looking for opportunities to learn and grow. This can be done by networking, reading articles, or publishing books. As long as you are learning something new, you will continue to grow. (pg. 386)
2. By mastering the use of your time through setting goals and planning your daily activities, you learn to control your time. Good time management skills require that you set priorities and discipline yourself to follow them but remain flexible to adapt to schedule changes if necessary. (pg. 383)
3. Look for individuals within your organization who could be developed into the next generation of instructors, and begin the development process with them. (pg. 390)

Instructor Applications

Answers to these questions will be based on personal experiences and local resources.